TOPICS IN PLANT PHYSIOLOGY: 3

SERIES EDITORS: M. BLACK & J. CHAPMAN

PLANT DEVELOPMENT

TITLES OF RELATED INTEREST

Basic growth analysis
R. Hunt

Class experiments in plant physiology
H. Meidner (editor)

Comparative plant ecology
J. P. Grime, J. G. Hodgson & R. Hunt

Crop genetic resources
J. T. Williams & J. H. W. Holden (editors)

Introduction to vegetation analysis
D. R. Causton

Introduction to world vegetation (2nd edition)
A. S. Collinson

Light & plant growth
J. W. Hart

Lipids in plants and microbes
J. R. Harwood & N. J. Russell

Plant breeding systems
A. J. Richards

Physiology & biochemistry of plant cell walls
C. Brett & K. Waldron

Plants for arid lands
G. E. Wickens, J. R. Goodin & D. V. Field (editors)

Plant tropisms
J. W. Hart

PLANT DEVELOPMENT

THE CELLULAR BASIS

R. F. Lyndon

Department of Botany
University of Edinburgh

London
UNWIN HYMAN
Boston   Sydney   Wellington

Published by the Academic Division of
Unwin Hyman Ltd
15/17 Broadwick Street, London W1V 1FP, UK

Unwin Hyman Inc.,
8 Winchester Place, Winchester, Mass. 01890, USA

Allen & Unwin (Australia) Ltd,
8 Napier Street, North Sydney, NSW 2060, Australia

Allen & Unwin (New Zealand) Ltd in association with the
Port Nicholson Press Ltd,
Compusales Building, 75 Ghuznee Street, Wellington 1, New Zealand

First published in 1990

British Library Cataloguing in Publication Data

Lyndon, R. F.
  Plant development.
1. Plants development
I. Title II. Series
581,3
ISBN 0–04–581032–X
ISBN 0–04–581033–8 Pbk

Library of Congress Cataloging-in-Publication Data

Lyndon, R. F.
  Plant development: the cellular basis/R. F. Lyndon.
    p. cm. – (Topics in plant physiology; 3)
Includes bibliographical references.
ISBN 0–04–581032–X (alk. paper). – ISBN 0–04–581033–8 (pbk.
alk. paper)
  1. Plants – Development. 2. Plant cells and tissues. 3. Plant
molecular genetics. I. Title. II. Series.
QK731.L96 1990
581.3–dc20   90–24858
                                                      CIP

Typeset in 10 on 12 point Palatino by Computape (Pickering) Ltd,
North Yorkshire
and printed in Great Britain by Cambridge University Press

Preface

The study of plant development in recent years has often been concerned with the effects of the environment and the possible involvement of growth substances. The prevalent belief that plant growth substances are crucial to plant development has tended to obscure rather than to clarify the underlying cellular mechanisms of development. The aim in this book is to try to focus on what is currently known, and what needs to be known, in order to explain plant development in terms that allow further experimentation at the cellular and molecular levels. We need to know where and at what level in the cell or organ the critical processes controlling development occur. Then, we will be better able to understand how development is controlled by the genes, whether directly by the continual production of new gene transcripts or more indirectly by the genes merely defining self-regulating systems that then function autonomously.

This book is not a survey of the whole of plant development but is meant to concentrate on the possible component cellular and molecular processes involved. Consequently, a basic knowledge of plant structure is assumed. The facts of plant morphogenesis can be obtained from the books listed in the General Reading section at the end of Chapter 1. Although references are not cited specifically in the text, the key references for each section are denoted by superscript numbers and listed in the Notes section at the end of each chapter. From these it should be possible to gain access to the literature on each topic. Points in the text that do not seem to be covered in the key references will normally have been covered under Further Reading, which often includes books and symposia in which there are other papers of great relevance and interest but that are not cited individually. Botanical terms that the reader may find helpful to have explained are given in the Glossary.

Many people have been generous in giving their time to read and comment on parts of the manuscript. I am grateful to the editors, Mike Black and John Chapman, and Clem Earle of Unwin Hyman, for going carefully through the whole text. Without Clem's coaxing, this text may never have seen the light of day. I have very much appreciated the extremely helpful comments of Adrian Dyer, Dennis Francis, Mark Fricker, Paul Green, Ricardo Murphy, Tsvi Sachs, Steve Smith, Tony Trewavas and John Tulett. Although I have taken all their points into consideration I have, perhaps unwisely, not always chosen to act on their advice. I have also been very gratified by the spontaneous and generous help and comments of colleagues throughout the world when they were approached for permission to use published material.

R. F. Lyndon

Acknowledgements

We are grateful to the following individuals and organisations who have kindly given permission for the reproduction of copyright material (see References at the end of this book for full details of sources; figure numbers in parentheses):

B. E. S. Gunning (2.2); Figure 2.5 from Coe & Neuffer 1978, by permission of E. H. Coe and Academic Press Inc. (2.5); L. M. Blakely and the American Society of Plant Physiologists (3.1); A. Zobel and Annals of Botany Co. Ltd (3.2); K. E. Cockshull and Annals of Botany Co. Ltd (3.9); Figures 4.3 and 4.8 from Green 1984, by permission of P. B. Green and Academic Press Ltd (4.3, 4.8); Figure 4.4 from Hardham *et al.* 1980, by permission of P. B. Green and Springer–Verlag (4.4); P. B. Green (4.11); P. B. Green and MGR Publications (4.7); P. B. Green and the Company of Biologists Ltd (4.9, 4.10a); J. Jeremie (5.1); P. M. Lintilhac and Cambridge University Press (5.4); P. M. Lintilhac and MGR Publications (5.5, 5.6); Figure 5.8 from Palevitz & Hepler 1974, by permission of B. A. Palevitz and Springer–Verlag (5.8); Figure 5.10 from Palevitz 1981, by permission of B. A. Palevitz and Cambridge University Press (5.10); Figures 6.1, 6.3 and 6.4 from Weisenseel & Kicherer 1981, by permission of M. H. Weisenseel and Springer–Verlag (6.1, 6.3, 6.4); Figure 6.2 from Robinson & Jaffe 1975, by permission of L. F. Jaffe, © 1975 by the AAAS (6.2); Figure 6.9 from Robinson *et al.* 1984, by permission of J. H. Miller and Springer-Verlag (6.9); Botanical Society of Japan (6.10, 6.12); Figure 6.13 from Reiss & Herth 1978, by permission of W. Herth and Springer–Verlag (6.13); Figure 6.14 from Sievers & Schnepf 1981, by permission of A. Sievers and Springer–Verlag (6.14); Figures 6.15 and 6.17 from Kallio & Lehtonen 1981, by permission of Springer–Verlag (6.15, 6.17); Figure 6.16 from Kiermayer 1981, by permission of Springer–Verlag (6.16); Figure 6.18 from Meindl 1982, by permission of Springer–Verlag (6.18); MGR Publications (7.2); A. Komamine and the American Society of Plant Physiologists (7.3); A. Komamine and the Editor, *Physiologia Plantarum* (7.4); D. H. Northcote and the Company of Biologists Ltd (7.6); Figure 7.7 from Jacobs 1979, by permission of W. P. Jacobs and Cambridge University Press (7.7); Figure 7.10 from Sachs 1984, by permission of T. Sachs and Cambridge University Press (7.10); Figure 7.12 from Gersani & Sachs 1984, by permission of T. Sachs and Springer–Verlag (7.12); Figures 7.11 and 7.13 from Sachs 1981, by permission of T. Sachs and Academic Press

Inc. (7.11, 7.13); Figure 8.1 from Barlow 1976, by permission of P. W. Barlow and Academic Press Ltd (8.1); Figure 8.3 from Heyes & Brown 1965, by permission of J. K. Heyes and Springer–Verlag (8.3); R. O. Erickson (8.4, 8.5); Figure 8.6 from Erickson 1966, by permission of R. O. Erickson and Oxford University Press (8.6); R. Leech and the American Society of Plant Physiologists (8.7); Figure 8.8 from Roberts *et al.* 1985, by permission of K. Roberts and Oxford University Press (8.8); P. W. Barlow and Annals of Botany Co. Ltd (8.10); J. Van't Hof and MGR Publications (8.11); Figure 9.1 from Dure 1985, by permission of L. Dure and Oxford University Press (9.1); R. B. Goldberg and the Editor, *Cell* (9.2); Figure 9.3 originally published in Philosophical Transactions of the Royal Society, Series B, volume 314, reproduced by permission of D. C. Baulcombe and the Royal Society (9.3); Figure 10.1 from Walker *et al.* 1979, by permission of Elsevier Scientific Publishers Ireland Ltd (10.1); Figure 10.2 from Christianson & Warnick 1985, by permission of M. L. Christianson and Academic Press Ltd (10.2); Figure 10.4 from Meins & Binns 1979, by permission of F. Meins, © 1979 by the American Institute of Biological Sciences (10.4); E. G. Williams and Annals of Botany Co. Ltd (10.6); J. Warren Wilson and Annals of Botany Co. Ltd (10.8); H. Kende and the American Society of Plant Physiologists (10.9); T. Sachs and the Editor, *Israel Journal of Botany* (10.10); Figure 11.2 from Kinet *et al.* 1971, by permission of J. M. Kinet, G. Bernier and Gustav Fischer Verlag (11.2); Figure 11.3 from Dennin & McDaniel (1985), by permission of C. N. McDaniel and Academic Press Ltd (11.3); R. W. King, L. T. Evans and the Editor, *Australian Journal of Biological Science* (11.4); E. Miginiac and the Editor, *Plant Physiology and Biochemistry* (11.5); J. Brulfert (11.7); Figure 12.2 from Esau 1965, © 1953, 1965 John Wiley & Sons Inc., by permission of K. Esau and John Wiley & Sons Inc. (12.2); Figures 12.3 and 12.4 from Barlow & Adam 1988, by permission of P. W. Barlow and Springer–Verlag (12.3, 12.4); A. Lindenmayer and Alan R. Liss, Inc. (12.6a); Figure 12.6b from Cannell 1974, by permission of M. G. R. Cannell and Blackwell Scientific Publications Ltd (12.6b); Figures 12.9 and 12.10 from Marx & Sachs 1977, by permission of T. Sachs and The University of Chicago Press (12.9, 12.10); Figure 12.11 from Sachs 1974, by permission of T. Sachs and The University of Chicago Press (12.11); Figure 12.14 from Warren Wilson 1984, by permission of J. Warren Wilson and Cambridge University Press (12.14); Figures 12.15 and 12.16 from Meinhardt 1984, by permission of H. Meinhardt and Cambridge University Press (12.15, 12.16); Figure 12.18 from Harrison *et al.* 1981, by permission of L. G. Harrison and Springer–Verlag (12.18); L. G. Harrison and the Editor, *Canadian Journal of Botany* (12.19); D. G. Mann and Koeltz Scientific Books (12.20).

Contents

CONTENTS

List of special topic boxes

List of tables

Abbreviations

2, 4-D	2, 4-dichlorophenoxyacetic acid
2-D	2-dimensional
2C, 4C	2, 4 times the gametic (C) amount of DNA
$2n$, $4n$, $64n$	2, 4, 64 times n, where n is the haploid chromosome number
3-D	3-dimensional
ABA	abscisic acid
BM	basal medium
bp	base pairs
CB	cytochalasin B
cDNA	complementary DNA
CIM	callus-inducing medium
CYT	cytokinin
EDTA	Ethylene diamine tetraacetic acid
ER	endoplasmic reticulum
G_1	pre-S portion of interphase of the cell cycle (see S)
G_2	post-S portion of interphase of the cell cycle (see S)
GA	gibberellin
GA_3	gibberellic acid
GMC	guard mother cells (of stomata)
IAA	indol-3-yl acetic acid
IEDC	induced embryogenic determined cells
kb	kilobases
LD	long days
mRNA	messenger RNA
MDMP	2-(4-methyl-2, 6-dinitroanilino) N-methyl propionamide
MTOCs	microtubule organizing centres
MTs	microtubules
NPA	naphthyl phthalamic acid
PAGE	polyacrylamide gel electrophoresis
PAL	phenylalanine ammonia lyase
PCIB	p-chlorophenoxyisobutyric acid (clofibric acid)
PEDC	pre-embryogenic determined cells
PG	polygalacturonase
PPB	preprophase band (of microtubules)
QC	quiescent centre
RIM	root-inducing medium
RNase	Ribonuclease

rRNA	ribosomal RNA
S	DNA synthesis phase of the cell cycle
SD	short days
SIM	shoot-inducing medium
TIBA	triiodobenzoic acid
UDP	uridine diphosphate

PART I

Development of the basic structures

CHAPTER ONE

The problems of development: embryogenesis

1.1 THE PROBLEMS OF DEVELOPMENT

All sexual organisms begin as a single cell – the fertilized egg or zygote – which, by growth and **cell division**, ultimately gives rise to the whole mature organism. Because these cell divisions are all mitotic, every cell in the mature organism should have the same genetic constitution as the zygote. Despite this, cells differentiate into different types of cells, **tissues**, and organs (see the General Reading section at the end of the chapter). Not only do cells develop along different pathways, they do so only in particular places in the organism. The different cell types, tissues, and organs bear characteristic spatial relationships to each other. This three-dimensional form is specified in some way by the genes, since there are genes for leaf shape, for example, and the form of the flower is used by taxonomists as the most reliable index of the genetic differences between species.

The problem of development and differentiation is, therefore, how is gene expression modified or changed to produce different cell, tissue, and organ types? How are genes expressed selectively, not only at different times during development (i.e. sequentially activated or repressed) but also in different places within the organism? And how do genes which ultimately regulate the synthesis of proteins and which as enzymes determine only the rates of processes, govern the generation of three-dimensional form? In other words, how is information specifying the rates of processes translated into changes in shape? Or, to put this another way: how is the scalar information of the DNA transformed into the vectorial processes involved in the construction of a three-dimensional multicellular organism? The implication is that we need to be able to explain directions and planes of growth in terms of the rates of the underlying metabolic processes.

3

1.2   CHANGES IN GENE EXPRESSION

1.2.1   Genes are potentially active

Selective gene activity could be because genes are selectively deactivated or lost during development so that the developmental potential of different cells becomes progressively less in different ways. Nuclear transfer experiments in animals (Amphibia) suggest that nuclei, even from quite highly differentiated cells such as those of the gut, can sometimes produce a whole new organism if they are transplanted into an egg whose original nucleus has been removed, showing that genes are inactivated rather than lost or irrevocably damaged. In plants, this is seen much more easily because cells from differentiated tissues, such as **leaf mesophyll**, **epidermis**, secondary **phloem** (admittedly rather undifferentiated phloem), **pith**, **cambium**, petals, **ovules**, and **nucellus**, have all been made to regenerate whole plants in culture. The plant cell, at least while it still has a nucleus (which is lost in differentiation of some **xylem** and phloem cells), is therefore **totipotent**, i.e. capable of producing all other cell types and so re-creating the whole organism under the right conditions. During development, the genes must therefore be activated or repressed selectively (but not lost), so that as the organism develops only some genes are active at any particular time and in any particular part of the organism.

1.2.2   Genes are switched on at specific times and places and show scalar and vectorial expression

Molecular techniques have shown that genes for processes occurring during fruit ripening are switched on only during ripening. The transfer of genes from one species of plant to another by modern techniques has shown that genes coding for seed proteins are still expressed in the genetically engineered transgenic plants, so there must be controls of gene activity during development that are common to different species (see Ch. 9). But the control of development is presumably more complex than the switching on or off of single genes from time to time. Genes controlling the processes of cell division are inactive in non-dividing cells. Here, there seems to be a whole group of genes whose activity is coordinately switched on and off together. Even more complex examples are where genes controlling shape are switched on or off as organs are initiated and then begin to mature. Another problem is the scalar control of gene expression; genes are not necessarily just switched on or off, but may be switched on to a very small extent and then up-regulated, or if they are being actively expressed they may become down-regulated. Evidence is now becoming available (Ch. 9) of the sort of controls at the molecular level that may achieve this. But this pushes the problem one stage further back – how are the regulator genes themselves regulated and in a coordinated manner? A major problem of development is to explain how genes are expressed vectorially to specify the shape of an organ or organism. At the molecular level, we can begin to see how activities of individual genes or groups of genes may be regulated. At the other extreme, we can see what happens to the whole organism as it develops. In between, we need to see what is happening at the cellular level, which can explain the organization of the tissues, the organs, and the whole plant, and which in turn has to be explained by the changes in gene expression. It is this 'in-between' area that this book is about.

1.3   EMBRYOGENESIS – THE PROBLEMS EXEMPLIFIED

Many of the problems of development are highlighted by the processes occurring during embryogenesis. The development of the embryo of *Capsella* (Shepherd's Purse) is often taken as typical (Fig. 1.1).

1.3.1   *Polarity of the zygote*[1]

The entire embryo arises by growth and cell division from the fertilized egg, which is called the zygote. In seed plants, the egg, and consequently the zygote, is in a polar environment, i.e. the environments of the two ends of the egg differ, one end being against the embryo sac wall next to the **micropyle**, where the pollen tube enters, and the other end projecting into the fluid-filled embryo sac, containing high concentrations of growth substances and other metabolites.

The zygote itself displays polarity, i.e. it has an axis with the ends being different. It has a large vacuole at the micropylar end and dense cytoplasm, with the nucleus, at the embryo sac end. This is an accentuation of the polarity that already existed in the unfertilized egg, so it is not the result of the position at which the fertilizing nucleus enters – from the pollen tube – which is what would be expected by analogy with animal eggs. Whether the polarity of the egg results from the polarity of its environment in the embryo sac is not known. The polarity of the egg is translated into the structure of the embryo because the first division of the zygote is normally transverse, cutting off a densely cytoplasmic terminal cell, that gives rise to the embryo proper, and a highly vacuolate basal cell that gives rise to the suspensor. In the embryos of some species, the first division is longitudinal; whether this corresponds to a lateral polarization of the zygote is not known, but it seems doubtful.

Evidence that the polarity of the zygote and the subsequent embryo is the result of the physical constraints of its immediate environment come

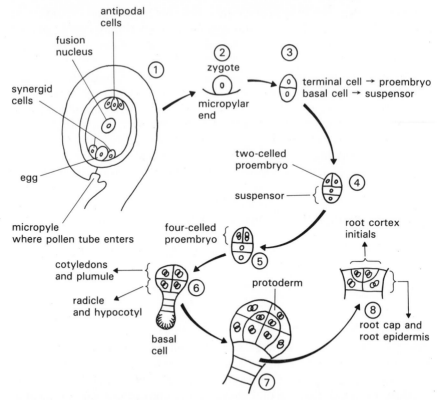

Figure 1.1 Early development of *Capsella* embryo. The male nucleus (haploid, *n*) from the pollen tube fuses with the egg to give the zygote (2*n*). The vegetative nucleus (*n*) from the pollen tube fuses with the fusion nucleus (2*n*) to give a triploid nucleus (3*n*), which divides to give rise to the endosperm cells (3*n*).

from experiments with the ferns *Phlebodium* and *Thelypteris*, in which the zygote was severed by cuts from the surrounding **archegonium** walls, and *Todea*, in which the embryo was isolated and cultured. The embryo then developed as a somewhat unordered, leaf-like, and much-branched structure, which only later became organized to form a young **sporophyte**. Supplying auxin could, in some cases, restore normal development. **Angiosperm** embryos are probably under pressure from the liquid contents of the embryo sac. In embryogenesis in culture (*in vitro*), ordered development seems to be promoted by a relatively high concentration of solutes in the medium, which mimics the high concentration found naturally in the embryo sac. The high pressures thus exerted on developing embryos by physical restraint and pressure may help to prevent premature cell expansion, physically controlling the growth of the embryo and so bringing about the precise orientation of the planes of cell division within it.

Polarity can arise within the embryo itself in free-living embryos, as in the seaweed *Fucus*, as a result of many different external stimuli (see Ch. 6, Table 6.1). Embryos can, however, form and develop in an apparently apolar environment, as shown by the development of embryos from zygotes of the alga *Homosira* in shake culture and from continually agitated angiosperm cell suspension cultures. In angiosperm cell suspensions, where the cells are presumably in an apolar environment, the cells nevertheless usually, or always, become polarized by the development of a long axis. Whether this depends on being triggered by some external stimulus (however fleeting) or whether it develops spontaneously within the cells, and whether the development of polarity is essential for embryo development or can arise as a consequence of it, is not known. Morphological and physiological polarity is characteristic of all plants and can be studied at the cellular level. Only when cells form unorganized callus is polarity absent.

1.3.2 *Formation of embryo and suspensor – a division of labour?*[2]

The first cell division in the embryo gives the terminal cell, which develops into the **proembryo**, or embryo proper, and the basal cell, which gives rise to the suspensor. (The division of labour is not always as clear cut as this, e.g. in *Capsella* some derivatives of the basal cell give rise to part of the embryo; see Fig. 1.1.) The elongating suspensor pushes the developing embryo into the embryo sac and the **endosperm**, which is a metabolite source for the embryo. The polarity of the embryo's environment is therefore maintained because one end of the embryo is free in the embryo sac or endosperm and the other is attached to the suspensor. Is the role of the suspensor to maintain the embryo in a polarized environment, or does it have other functions as well?

It seems an attractive idea that the polarity of the embryo might be partly the result of the opposing sources of nutrients that presumably result from a supply from the embryo sac at one end and from the suspensor at the other. However, this is unlikely. The zygote and the young embryo are covered by a **cuticle**, suggesting that the surface in contact with the embryo sac and endosperm may not (in the early stages, anyway) be an absorptive surface, but that the preferred transport channels to the embryo are via the suspensor. Also, the basal cell of the suspensor soon develops into a **transfer cell** (see Ch. 8, section 8.3.1.1), suggesting that there is an important flux of material into and through it. In *Phaseolus*, the suspensor is relatively large and can be dissected out, with its young embryo attached, in an apparently undamaged state. Up to the heart stage of development the suspensor acts as a supplier of substances to the embryo, but once the **cotyledons** have formed they take over as the major absorptive organs for the embryo (Table 1.1). Before the

7

Table 1.1 Uptake of [^{14}C]-sucrose by suspensor and embryo of *Phaseolus vulgaris*. In the younger embryos (late heart stage), the suspensor is the most efficient absorbing part of the embryo, as shown by the uptake ratio; in the older embryos (mid-maturation), the cotyledons become the most efficient absorbing region. [^{14}C]-sucrose was injected into the base of the pod. Embryos were dissected out and analysed 7 h later. Similar results were obtained when injection was into the endosperm cavity around the embryo. (Yeung 1980)

Stage of embryo development	Embryo region	% Counts (A)	% Fresh wt (B)	Uptake ratio (A/B)
late heart	suspensor	21.7	7.1	3.05
	proembryo half next to suspensor	44.7	40.4	1.11
	proembryo half away from suspensor	33.8	52.4	0.65
mid-maturation	suspensor	0.08	0.12	0.64
	embryonic: axis	0.7	1.5	0.43
	base	19.8	24.6	0.80
	cotyledon: mid	46.6	46.3	1.01
	tip	31.5	27.6	1.14

major axis and main regions of the embryo have been mapped out, the embryo probably depends mostly on the suspensor for its supply of nutrients.

The suspensor may not be essential, for suspensor-less adventive embryos in culture grow successfully as they also do in *Citrus*, which shows polyembryony and in which only the zygotic embryo has a suspensor and the accessory asexually derived embryos have not. The suspensor does not seem to have a role in embryo development, except perhaps to supply the young embryo with growth substances and to push it into the nutrient-rich medium in the embryo sac. It may contribute gibberellin to the developing embryo up to the heart stage, since lack of a suspensor in young cultured embryos can be compensated for by a supply of gibberellin in the culture medium (Table 1.2). Transfer of materials from the suspensor to the embryo does not only occur when they are in organic contact – placing them next to each other in culture is also effective. Continuity of the **symplast** of the suspensor and embryo via the **plasmodesmata** linking them may normally facilitate development but there is no symplasmic contact of the basal cell with the embryo sac. The base of the basal cell may have no plasmodesmata but develops instead as a transfer cell. The lack of plasmodesmata here also implies that the embryo and its suspensor are an isolated symplast entity within

Table 1.2 Effect of gibberellic acid (GA_3) on embryo growth of *Phaseolus coccineus*. Older (longer) embryos can grow in culture without a suspensor. Younger embryos need a suspensor but it can be replaced by a supply of gibberellin. (Cionini *et al.* 1976)

Embryo length (mm)	% Embryos grown into plantlets after 15 d in culture		
	Intact embryo	Embryo minus suspensor	Embryo minus suspensor + 10^{-8}M GA_3
0.5	14	4	17
1.0	39	13	24
1.5	65	36	68
2.0	89	37	59
3.0	100	66	89
5.0	100	98	–

the parent plant. Whether this isolation is necessary for embryo development is a point that is still debated. Nevertheless, the embryo and suspensor in the normal plant are an example of differentiation, from adjacent sister cells of the same parent cell, of two groups of cells that complement each other in their growth and functions, even though it may be mostly a one-sided complementation with the embryo contributing little or nothing to the suspensor.

1.3.3   Rates and planes of cell division[3]

The cells in the suspensor do not increase in number as fast as in the proembryo and are limited in number, unlike the proembryo in which cell number is potentially indefinite as it grows into the adult plant. The rate and duration of cell division therefore differ in the proembryo and suspensor. In a mutant of *Arabidopsis* in which the young globular embryo ceases growing and aborts, the suspensor grows on to form a chain of two or more files of 15–150 cells. The growth potential of the suspensor is therefore revealed only when the inhibitory action of the proembryo is removed. Whether embryos in general have this inhibitory effect on the suspensor, and whether this is the explanation for the different division rates in adjacent tissues, is not known.

As the embryo develops it also changes in shape from globular to heart shaped when the cotyledons are formed. Cell divisions are apparently more frequent in the developing cotyledons than in the centre of the embryo. After the first division of the zygote, usually transverse as in *Capsella*, the next division in the terminal cell is usually in a different plane, normal to the first and therefore longitudinal (Fig. 1.1). However, in the basal cell, which forms the suspensor, the second and subsequent divisions are all in the same plane as the first and are therefore transverse. The terminal group of cells goes on to divide in other planes to form a mass of cells, called the globular embryo, which is anchored to the embryo sac wall by the file of cells that is the suspensor (Fig. 1.1). The various planes of division in the proembryo reflect its generally isotropic (equal in all directions) growth, and the transverse divisions in the suspensor reflect its linear growth. The problem is: do the directions of growth dictate the planes of division or do the planes of division dictate the direction of growth? In some embryos the suspensor is elongated, as usual, but is noncellular and multinucleate (e.g. *Orobus* or *Pisum*, in which there are several such cells). This suggests that the plane of cell division is probably the result of the general direction of growth and not its cause (see also Ch. 5).

In the *Capsella* proembryo the successive cell divisions are in precise and predictable planes, at least for the first five or so divisions, after which divisions are harder to follow. What causes the divisions to be orientated so precisely? If the overall shape of the globular embryo is the result of controls on the embryo as a whole rather than on each individual cell, then the positions of the cell walls might simply be those that are the consequence of physical or structural restraints in the embryo. The partitioning of a cell and its daughters according to simple rules could conceivably produce the pattern seen in globular embryos. The actual algorithm operated by the *Capsella* embryo seems to specify that each division is normal to the previous one until the formation of the protoderm (the embryo epidermis), which is formed at the 16-cell stage (Fig. 1.1) and which thereafter divides only anticlinally (i.e. divisions normal to the surface) until the mature embryo or seedling is formed.

In the suspensor, however, divisions are all transverse, giving rise to a string of cells one cell wide in *Capsella*. The plane of division therefore alters in successive divisions in the proembryo but remains constant in the suspensor. The control of the plane of division is therefore intimately linked not only to the shape of the resulting tissue (or organ) mass but also to its developmental fate. The changes in division planes in the embryo pose the broader question: to what degree is the shape of an organ caused by differential planes of division rather than by rates of growth? Is the plane of cell division a cause or a consequence of organ shape? Is the growth rate in **meristems** controlled by the cell division cycle or is the **cell cycle** controlled by the general rate of growth? These are questions that will be considered in later chapters (see Chs 5 & 2).

Since the rates and planes of cell division are obviously involved in the generation of plant form, the cell wall is a commensurately important organelle. In the plant, the walls not only encapsulate the developmental history of the organism but also provide the permanent structural framework that is the basis for all subsequent growth. Since plant form is inherited and is therefore the result of gene action, gene activity is expressed partly in the positions and planes of cell walls which depend on the previous orientation of the mitotic spindle. Three-dimensional cellular architecture, wall positioning, and the differential properties of the cell wall *within the same cell* require explanation in terms of gene action. The plane of cell division is ultimately the result of gene action, but how?

1.3.4 *Determination and differentiation*

The cotyledons form from the upper part of the embryo but, more than that, the lineage of the cells from which they are formed can be traced back to the 8-celled stage of the proembryo, and their developmental fate can be predicted (Fig. 1.1). This can be done in *Capsella* because of the regular pattern of cell division in the young embryo and the visible persistence of the cell wall boundaries between the different parts of the embryo as it grows. This is equivalent to the 'fate maps' constructed for animal embryos. The implication is that the fate of the cells is predictable at the 8-celled stage according to their positions in the embryo, and so the embryos show mosaic development. But this does not necessarily mean that the fate of the cells has become *fixed* at this stage. Even if the cells were visibly different in the early embryo, this would not necessarily imply that their developmental fates are fixed at this point, any more than they are in animal embryos. Whether or not the cells are determined, i.e. fixed in their fate, can only be measured by experiments that can cause them to follow some different fate if they are not yet determined.

1.3.4.1 **When do cells become determined?**
Determination is the process whereby the developmental fate of cells becomes fixed. Before determination the cell, tissue, or organ can be diverted into other developmental pathways, but once determined its fate is sealed and it becomes impossible to alter the developmental pathway on which the cells have embarked. The only way to find out whether the cells have become determined is to try to provoke the cells to develop in a different manner by placing them under different experimental conditions. If they cannot be made to develop in a different way, no matter how they are treated, they are said to be determined. Tests like this have been made with animal embryos that are amenable to experimentation because the eggs are laid external to the organism and are fertilized outside the animal. Amphibian embryos have been favoured because all the cells of the embryo contain yolk and therefore are not dependent on other cells for their continued nourishment. Such experi-

11

ments have shown that the first thing that becomes determined is the regional pattern (i.e. the differentiation into ectoderm, mesoderm, and endoderm), and tissue transplantation has shown that determination becomes progressive so that the organ systems and finally the cell types become determined.

The development of plant embryos seems broadly similar, although experiments comparable with those performed with amphibian embryos have not been done because of the inaccessibility of the plant embryo, enclosed as it is in an ovule (or, in the lower plants, an archegonium). It has proved difficult to dissect out young globular embryos that will recover and grow intact, and microsurgery does not seem to have been attempted. However, older embryos have been cut up and the consistent finding is that only the shoot portion will regenerate shoots, the cotyledon portion produces only callus, and the root and **hypocotyl** portions produce callus and roots. The cells are therefore determined as root or shoot (and possibly cotyledon) and epidermis, **cortex**, or **stele** by the time the seed has formed. Because plant cells show much greater plasticity than animal cells, the whole embryo can be separated into its component cells and each can apparently give rise to a whole new embryo and ultimately to a whole plant. Because even mature plant cells can often be made to dedifferentiate and grow into whole plants, plant cells (except those that lose their nucleus in the course of differentiation) are less fixedly determined than animal cells. Nevertheless, they do become determined until they are placed in conditions that allow dedifferentiation and redevelopment into a whole plant, although sometimes callus can then differentiate only into certain kinds of organs, e.g. roots or shoots (see Ch. 10).

We simply do not know whether the development of particular cells of an 8-celled plant embryo into root or shoot (Fig. 1.1) is because they are already determined or, because of their position, they necessarily end up in the root or the shoot and become determined later. The same applies at the 16-cell stage when periclinal divisions (parallel to the surface) have formed the protoderm – the embryonic epidermis. Although the epidermis remains morphologically distinct from this moment onwards, we do not know when the cells actually become determined. The plant epidermis, in becoming distinct early in development, is therefore comparable to an organ system in animal embryos rather than to the animal epidermis, which can originate from different organ systems of the animal embryo.

The early development of the plant embryo up to the establishment of the root-shoot axis and the cotyledon(s), and the formation of the provascular tissue, seems to resemble the development of the animal embryo in which cell fate depends on position in the embryo. In the plant embryo, the factors that position the cell walls and dictate the planes and

12

pattern of division may be superimposed on, but not necessarily cause, determination. The cells would become restricted in their eventual fate by virtue of the positions into which they become locked. The same may be true for animal embryos, in which cell movement only occurs later in development. Ultrastructural differences between cells (except for differences in ribosome concentration) have not been found in young plant proembryos, and presumably the factors causing determination are at the molecular level.

Polarity in the young embryo is shown morphologically by the establishment of the root–shoot axis, and physiologically by the polarity of auxin transport (see Ch. 7, Box 7.1), which can be demonstrated in the embryo before the seeds are ripe in *Phaseolus* (see Ch. 10, section 10.1). But it is not clear whether morphological polarity precedes polarity of auxin transport or whether polar auxin transport becomes established in the globular embryo and determines the morphological polarity. Calcium and auxin each promote the transport of the other, but whether polar auxin transport is established in the young embryo as a result of a calcium gradient, of the sort formed in *Fucus* zygotes (see Ch. 6, section 6.1), is not known. Auxin transport is certainly polarized when the **procambium** is just beginning to differentiate into phloem and xylem, so it may be that the development of the procambium into **vascular tissue**, or even the establishment of the axis of the procambium itself (see Ch. 7, section 7.6.2), could depend on a polar auxin gradient being established first in the young embryo.

1.3.5   Cell-to-cell communication

The development of cells in different parts of the embryo into differing tissue and organ systems (epidermis/cortex/stele/root/shoot/ cotyledons) poses the question of the nature of the positional information and how a particular cell recognizes where it is in relation to the positions of other cells (see Ch. 12). Presumably, there is some way in which information about the position of a cell relative to its neighbours can be transmitted from cell to cell. Animal cells, which can move about, recognize each other by an immunological type of mechanism on the cell surfaces that come into contact. Since plant cells are immobile, the contact from cell to cell is either via the plasmodesmata and thus through the symplast, or by the secretion and recognition of molecules across the plasma membranes and the cell walls. Although some products of cell wall degradation have been shown to have physiological effects on plant cells, it is unknown whether such molecules might be involved in early development, or indeed whether they are involved at all in cell–cell recognition.

The young embryo becomes isolated symplastically from the mother plant when the zygote loses its plasmodesmatal connections with the embryo sac. Also, when embryos are separated into their constituent cells the integrated growth of the whole embryo is replaced by each cell forming a whole embryo by itself. It seems probable, therefore, although it has not been proved, that symplastic isolation is necessary or at least tolerated for development of the whole embryo, but that symplastic continuity must be maintained within the embryo itself.

1.3.6   *Continued growth and development of the embryo*

The young embryo contains the major organ systems (root/shoot/ cotyledons), but only subsequently in the older embryo or during germination are the other organs of the plant formed, notably leaves, **lateral roots**, **axillary buds**, flowers and reproductive structures, and the mature tissue systems (phloem, xylem, **sclerenchyma**, and other specialized tissue and cell types). The embryo is just the basic pattern, and not necessarily even the whole of that. It is just the basic outline, which is later added to and filled in. The rest of the plant is formed by iterative development – the production of repeating units of structure by the meristems (see Ch. 2). Sequential development, as occurs in the embryo, only occurs later during formation of the flower when the shoot apex reverts, as it were, to an embryonic style of development in which new types of organs are formed. Embryogenesis in the ovule and fruit is equivalent to the whole of embryogenesis and development in the animal. The subsequent iterative development that is typical of plants has no counterpart in animals except perhaps in those lower animals that increase in size by budding. It is the iterative part of plant development that is susceptible to modification by the environment and where we may, therefore, expect to find control mechanisms that have no counterpart in animal development.

1.4   SUMMARY

(1)   Probably all nucleated plant cells are totipotent, i.e. they have the potential to develop into the whole plant. Differentiation is therefore the result of the restriction of gene expression in different ways in different cells, tissues, and organs.

(2)   The zygote, which is in a polar environment in the embryo sac, is itself polar and divides into a terminal cell that gives rise to the embryo proper and a basal cell that forms the suspensor. Embryos can also form in apparently apolar environments, so polarity may be intrinsic to the zygote and not have to be imposed from outside. Physical pressure, from the surrounding tissues or from high osmotic concentration in the surrounding fluids, may be necessary for organized development.

(3)   The suspensor's development is limited, perhaps by the proembryo, but it apparently assists in the transfer of nutrients, metabolites, and growth substances (probably gibberellin) to the proembryo. Cells of the suspensor, especially the basal cell, develop as transfer cells.

(4)   The plane of division changes in an ordered manner in the young proembryo of *Capsella* (although, in many species, early development is not so precise). In the suspensor, all divisions are transverse. Gene activity during development is, therefore, expressed partly in terms of positions and planes of cell divisions.

(5)   The ultimate fate of the different cells of the proembryo depends on their position only because cells at the top of the embryo become the shoot and cells at the base become the root; however, this does not necessarily mean that their fates have become fixed. This can only be found by experiment. The cells have become determined, because their development is not altered when the different parts of the embryo are isolated, by the time the embryo is fully formed.

(6)   The embryo, with its suspensor, is apparently isolated symplasmically from the mother plant because there are no plasmodesmatal connections between the embryo and the cells surrounding it. This suggests that symplasmic isolation may be a necessary precondition for cells to develop into embryos.

(7)   Embryogenesis in plants is comparable to the whole of development in animals. In plants, most development takes place after embryogenesis and is the result of iterative development at the meristems. It is meristematic activity and the resulting development that is most susceptible to modification by the environment.

GENERAL READING

Burgess, J. 1985. *An introduction to plant cell development*. Cambridge: Cambridge University Press. (Development of plant cell structure and ultrastructure)

Cutter, E. G. 1971. *Plant anatomy: experiment and interpretation. Part 2. Organs.* London: Edward Arnold. (Development of plant structure and morphology)

Cutter, E. G. 1978. *Plant anatomy. Part I. Cells and tissues*, 2nd edn. London: Edward Arnold. (Cell types and their differentiation)

Sinnott, E. W. 1960. *Plant morphogenesis*. New York: McGraw-Hill. (A valuable survey of the phenomena of plant development and how it can be modified by some external factors)

Wareing, P. F. & I. D. J. Phillips 1981. *Growth and differentiation in plants*, 3rd edn. Oxford: Pergamon Press. (Plant development, especially the action of applied growth regulators)

FURTHER READING

Esau, K. 1965 *Plant anatomy*, 2nd edn. New York: Wiley. (Plant structure and anatomy, including embryogenesis briefly)

Jensen, W. A. 1976. The role of cell division in angiosperm embryology. In *Cell division in higher plants*, M. M. Yeoman (ed.), 391–405. London: Academic Press. (Embryogenesis in cotton)

Johri, B. M. (ed.) 1984. *Embryology of angiosperms*. Berlin: Springer. (A valuable compendium of information about all aspects of embryogenesis in angiosperms)

Raghavan, V. 1976. *Experimental embryogenesis in vascular plants*. London: Academic Press. (Useful source book on embryogenesis)

Raghavan, V. 1986. *Embryogenesis in Angiosperms*. Cambridge: Cambridge University Press. (Embryogenesis, especially in culture)

Slack, J. M. W. 1983. *From egg to embryo*. Cambridge: Cambridge University Press. (Animal embryology, and concepts of development and determination)

NOTES

1 DeMaggio (1982), Raghavan & Srivastava (1982), Sussex, (1967). (Control of polar growth of embryos)
2 Bohdanowicz (1987), Cionini *et al.* (1976), Yeung (1980), Yeung & Clutter (1978), Yeung & Sussex (1979). (Role of the suspensor)
3 Meinke (1986). (*Arabidopsis*)

PART II

Iterative growth: Meristem structure and functioning

CHAPTER TWO

Root and shoot meristems: structure and growth

2.1 MODULE PRODUCTION

Plants, unlike animals, continue to develop throughout their lives. New structures are formed in a repetitive manner so that the plant has a modular construction. The module of the shoot is a stem **internode** and **node**, with a leaf and an axillary bud. The module of the root is a length of root bearing a lateral root. The plant continually increases in size by the addition of extra modules (also called phytomers) (Fig. 2.1). In order to describe the structure of almost the whole plant we only need to know the structure of one module from the shoot and one from the root, the total number of modules, and the branching pattern. On flowering, the new shoot modules are smaller and crowded and the lateral organs differentiate in new ways in the flower, but the basic modular construction persists until the pattern is ended by the formation of the male and female reproductive organs, the gametes, and ultimately, after fertilization, the production of a zygote, which forms a new embryo.

The parts of the plant that produce the iterated structures are the apical meristems. These are self-perpetuating regions of relatively undifferentiated dividing cells, which, as they grow, form the modules behind the tip. In trees, the shoot apical meristem continues to grow in this way for decades or even centuries. The root meristem is apparently simple because it functions in a constant manner, maintaining itself and producing only an axis on which lateral roots form several millimetres or centimetres behind the meristem. The shoot apex forms leaves or floral organs on the meristem itself and so its growth, unlike that of the root meristem, is obviously repetitive and cyclic. This implies some sort of local feedback, since the positions of new organs are predictable in relation to the positions of those already present. The operation of plant meristems and the form of the organs produced can also be affected to

19

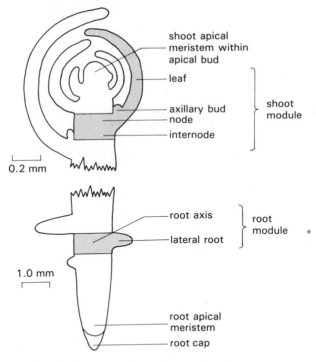

Figure 2.1 Modular construction of a plant. The apical meristems produce the same basic units of structure again and again. When axillary buds and lateral roots grow out, the pattern is repeated. The internodes elongate to form the stem.

varying degrees by the external environment. We may, therefore, perhaps expect to find control systems in plant meristems that do not have obvious counterparts in animals.

2.2 MERISTEM ORGANIZATION

2.2.1 *Meristems with apical cells*[1]

The organization of the meristem is most easily seen in those plants with apical cells, i.e. lower plants, including mosses, horsetails, and ferns. All cells are formed by division of the apical cell (Box 2.1). The root of the small aquatic fern *Azolla* has been studied in detail. There are only about 9000 cells in the mature root, and the clear cell wall pattern allows the lineage of any cell to be traced (Fig. 2.2). The tetrahedral apical cell divides first, to cut off from its **distal** face a cell that divides by one further parallel division and then numerous anticlinal divisions to give the root cap. All subsequent divisions of the apical cell are from its three basal

20

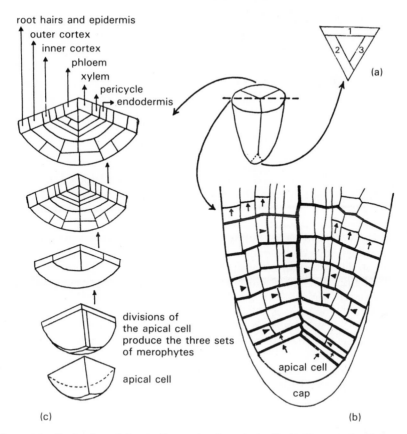

Figure 2.2 Production of the *Azolla* root by the apical cell. (a) The apical cell (here in transverse section) divides sequentially from its three basal faces to give the three sectors of the root. The sequence of division (1,2,3) can be inferred from the positions of the cell walls. (b) Longitudinal section of the root showing the merophytes (bounded by thick lines), the groups of cells each formed by a single derivative of the apical cell. Horizontal arrowheads show the longitudinal (formative) divisions, and the arrows show the proliferative divisions, which give the longitudinal files of cells. (c) Exploded view of one merophyte line, forming one sector of the root, showing how cell files and cell types arise. (After Gunning 1982)

faces in strict succession, to give the cells that make up the root proper. The derivative cells themselves divide (about 5–10 times in total) in a precisely ordered fashion (reminiscent of the ordered divisions in some embryos) to give the cells and tissues of the root (Fig. 2.2). The apical cell divides about 55 times before it ceases growth in the mature root. The rate of cell division in its immediate derivatives (about every 8–12 h) is slower than in the apical cell itself (about every 3–5 h). Although the fates of the derivatives of the apical cell can be mapped, it is not known when they become determined. The form of the root is not affected by the exact

21

sequence of divisions, which, as in embryogenesis, may not always be exactly the same. Apical cells in other plants appear to function in a similar way, although more divisions may be required to produce the root structure. Shoot apical cells function in the same way, except that no cells are cut off from the apical face and so there is no cap. Cell division is restricted to the three basal faces of the shoot apical cell.

Box 2.1   Apical cells[2]

As it enlarges and divides, the sequence of divisions in a tetrahedral apical cell (1, 2, 3 in Fig. 2.2) can be inferred from the cell wall pattern. In the root, cells are cut off from the 4th face to give rise to the root cap. The cytoskeleton is presumably involved in some way in determining the change in division plane, but how is not known. The cells subsequently formed by further divisions of an immediate derivative of the apical cell are recognizable as a group because of the division pattern and because of the slightly thicker cell wall surrounding them. An apical derivative, or the group of cells derived from it, is known as a merophyte (Fig. 2.2), which is analogous to the 'packets' of cells found in higher plant meristems. In those mosses (e.g. *Polytrichum*) in which each merophyte gives rise to a leaf, measurement of the rate of leaf production therefore gives the rate of division of the apical cell. The apical cell and its nucleus are usually much larger than other cells but the physiological basis for this is not known. Although some old apical cells may become polyploid and non-functional as initials, younger and functional apical cells are diploid. In ferns, a primordium can sometimes develop either as a leaf or a shoot. Once the apical cell has formed, the development can no longer be switched into the other pathway and so the primordium has become determined (see Ch. 10, section 10.8.2).

In **liverworts** and mosses with three ranks of leaves, these correspond to the three faces of the apical cell, each of which forms an identifiable sector of the plant. Apical cells sometimes cut off cells from only two sides, as in mosses with two ranks of leaves, each rank and half the plant being the product of one face of the apical cell. In other mosses and ferns where the leaves are not in ranks but in Fibonacci arrangement (see Ch. 3, section 3.2.2), there is no correspondence. If an apical cell is destroyed, a new one can be formed by one of its derivatives taking over its function. This implies that it is the position of the cell rather than its lineage that determines its structure and function, as in the higher plant meristem (see Box 2.2).

(a) (b)

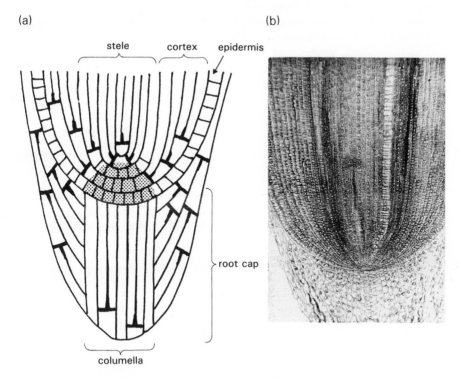

Figure 2.3 Initials in the maize root apex. (a) The stele, cortex, and epidermis can be traced back to layers of initial cells in the quiescent centre (shaded). Files of cells and the positions of longitudinal divisions are shown. Divisions transverse to the files are shown only for the QC, which can be deduced in this example from the presence of a columella and the lack of congruity of cell walls between initial layers, implying that there is minimal lateral growth in the QC region and no cell exchange between initial layers. (b) Metaxylem files can be traced back to the initials.

2.2.2   Meristems with multiple initials

2.2.2.1   The root meristem[3]

Root tips of higher plants do not have a single apical cell from which all other cells are derived. In those roots with closed meristems, the main tissue systems of the root can be traced back to only a few cells at the tip (Fig. 2.3). In the maize root meristem, the **initial cells** for the stele, cortex, epidermis, and root cap appear distinct so that there is a minimum of four cells acting as initials. The cell walls of adjacent layers are staggered, so the cells in adjacent initial layers do not look as though they are derived from each other. Each layer seems to have its own initial cells, which contribute cells, by division, to that layer only and not to adjacent layers. In each layer it is likely that there are at least three initials rather than a unique axial cell in each case. This would mean that a minimum of about

12 cells form the **promeristem**, which is self-perpetuating and performs the same function as an apical cell in giving rise to the rest of the root. This part of the root promeristem, which contains the epidermal, cortical, and stelar initials, grows and divides very slowly and so is called the quiescent centre (QC) (see section 2.3.1).

Roots with open meristems are those in which the initial layers for the cortex, epidermis, and root cap are not distinct, apparently because there may be a single set of initials for these layers. Temporary initials, which function like those in closed meristems, are therefore periodically replaced.

The shape of all roots is similar, broadening just behind the tip. This is where the formative longitudinal cell divisions are found, which increase the number of longitudinal cell files and form laterally adjacent cells that may differentiate into different cell types (see Fig. 2.2). The cells in these files divide transversely to increase root cell number as the root grows forward.

Cell differentiation is initiated in the meristem, and for the large metaxylem elements in maize it can be traced right back to the initials in the QC (Fig. 2.3). Quiescent centres can be isolated from the rest of the root as they are more resistant to disruption when the root is treated with EDTA, perhaps because of the structure of their cell walls. When placed on a suitable medium, an isolated QC will grow and form a root from its **proximal** acroscopic face, on which the pattern of the metaxylem is just visible. It also forms a root cap from its distal face. Since the root cap initials normally lie just outside the QC, this means that the QC must have produced new root cap initials, as it can do when the root is damaged. This shows that all the necessary cell initials and pattern-generating centres reside in the QC. These initial cells and tissue patterns become established in the embryo in the case of the primary root and in the new lateral meristems in the case of secondary roots. The newly emerged **radicle** of a germinated seed, and young secondary roots, do not have a QC but one develops as the root grows, showing that it is a consequence and not a cause of root growth and structure. The tissue pattern is first established in the embryo radicle and in each new root **primordium**, and can be modified experimentally (see Ch. 12, section 12.2). As cells are displaced away from the root initials, they differentriate and mature (see Ch. 8).

2.2.2.2 The shoot meristem[4]
When shoot apices with at least two young leaf primordia on them are excised and cultured on a suitable medium, but without added auxin and cytokinin, they can continue to grow and form leaves in the same arrangement as in the intact plant. The isolated **apical dome** itself, without any primordia, goes on to grow and produce new leaves only if

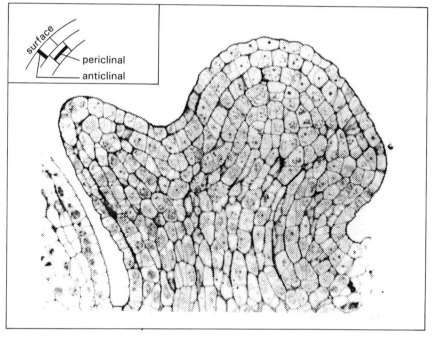

Figure 2.4 Initials in the pea shoot apex. The outer (tunica) layers have only anticlinal divisions. This layered structure is disrupted by periclinal divisions in the outer layers where a leaf is forming. (Lyndon & Cunninghame 1986)

cytokinin and auxin are provided, indicating that these are necessary for leaf initiation and that they are present in apices that already have young leaves. This suggests that there is a feedback loop operating in which auxin and cytokinin are produced by young leaves and are necessary for the initiation of further leaves. Cytokinin and auxin may, therefore, be needed to induce their own synthesis in newly initiated leaf primordia. These experiments also show that all the initial cells necessary to form a shoot are present in the shoot apical dome.

In the shoot meristem there is no cap and no layering of initials, except for the **tunica–corpus** structure (Fig. 2.4), which does not correspond to the tissues into which the cells differentiate. Indeed, in leaf formation there is a good deal of variability, even in a single plant, as to which cell layers of the apex give rise to which cell layers of the leaf. There may be no permanent initials in the shoot apex, just as there is no germ line (Box 2.2).

Box 2.2   Clonal analysis[5]

Chimeras are plants that are a composite of layers or sectors of cells of different genotypes. In some chimeras (e.g. *Cytisus adamsonii*), the layers are even from different species. Chimeras may arise naturally or can be produced artificially (e.g. the graft chimera of *Camellia*, see Ch. 3, section 3.2.1). Regions of a plant in which anthocyanin pigment or chloroplasts are absent can be distinguished easily wherever they occur. Occasional single mutant or defective cells can be induced in the shoot meristem by irradiation or treatment with chemical mutagens. Some of the treated plants can show whole sectors (as seen in plan view) that are different. If one-third of the plant is different this implies that there were $3/1=3$ initial cells at the time of treatment, if one-tenth then there were ten initials and so on. The vertical length of a sector reveals the number of plastochrons (the time for the number of leaves in it to be initiated) that the particular initial was acting as an initial before it was displaced from the meristem and replaced by another initial. Experiments like this have shown that the different parts of the maize plant can be traced back to relatively few shoot apical initials in the embryo (Fig. 2.5). In *Helianthus* (sunflower), mutant sectors extend through many internodes and leaves and continue up into the **inflorescence**, showing that the same shoot meristem initials can give rise to vegetative and reproductive parts of the plant (even though the histological evidence suggests that *Helianthus* shoot apical initials divide very slowly). Mutant sectors also usually end at a node and not in the middle of an internode, which is consistent with the internode being derived from a single layer of initial cells (see Ch. 3, section 3.2). In ferns, no phenotypic sectional chimeras have ever been described. This is consistent with all cells having a common origin, the apical cell, rather than arising from a group of initials as in the higher plants.

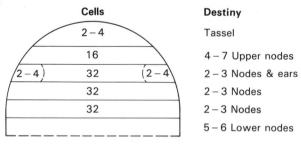

Cells			Destiny
	2 – 4		Tassel
	16		4 – 7 Upper nodes
2 – 4	32	2 – 4	2 – 3 Nodes & ears
	32		2 – 3 Nodes
	32		2 – 3 Nodes
			5 – 6 Lower nodes

Figure 2.5 Clonal analysis in maize. Shoot apical dome of the embryo showing the origin of different parts of the plant. Apices irradiated later, during germination, showed that specific cells could have various fates and must have become determined only after germination. (Coe & Neuffer 1978)

Clonal analysis is a very powerful tool for following cell lineages because very infrequent divisions of the apical initials can be detected. It has shown that shoot apical cells can function as initials for much of the life of the plant but that there are no permanent initials. The initial cells are those that happen to be at the summit of the meristem and are no different from other cells. It is their position that gives them their status as initials (see Ch. 12, section 12.4.1). There is also no specialized germ line in plants. This carries the implication that any environmentally induced changes to the genotype of any meristem cells have the possibility of being transmitted through the gametes to the next generation (see Ch. 9, section 9.3).

The shoot apex undergoes a major transition when it becomes floral. Its structure changes, its growth rate increases, it initiates new kinds of organs faster and in new arrangements (see Ch. 3, section 3.3), it forms the male and female organs in which meiosis uniquely occurs, and it becomes determinate in its growth.

2.3 HOW IS THE RATE OF CELL DIVISION CONTROLLED IN MERISTEMS?

2.3.1 Root meristems[6]

Unlike *Azolla*, in higher plant roots the rate of cell division in the promeristem and the initials is lower than in their derivatives (Table 2.1). This region of slowly dividing and growing cells at the centre of the meristem is called the quiescent centre (QC) (Fig. 2.3). It can be identified in autoradiographs of root sections by its lack of incorporation of [^{3}H] thymidine (a DNA precursor) because of its low rate of DNA synthesis. It is more or less hemispherical in shape in closed meristems, but in open meristems it is more disk-shaped. The faster dividing cells, including the initial cells of the root cap, are just outside the QC (Table 2.1). Because the cells of the QC grow and divide, however slowly, they are gradually displaced outward to become the growing and dividing cells of the meristem before they too are eventually displaced by cells from the QC. These dividing cells on the proximal face of the QC are analogous to the rapidly dividing proximal faces of the *Azolla* apical cell. The slowly growing distal face of the QC, abutting on the root cap, is analogous to the distal face of the *Azolla* apical cell, which expands only very slowly and does not produce any more cells after its first divisions.

In very young or small roots, there is no QC; it develops as the root ages and grows bigger. The QC is largest, and most marked, in those roots with closed meristems where there is a clear boundary and a thick cell wall between the epidermis, which is in the QC (Fig. 2.3), and the root cap initials outside it. When this root/root cap boundary consists of a wall that expands laterally in surface area only very slowly, this restricts

27

Table 2.1 Cell cycle in maize root tips. Lengths of the cell cycle and its component phases measured by pulse labelling with [^{3}H] thymidine in *Zea mays*. C = total cycle; S = DNA synthesis; M = mitosis; G_1 = pre-S interphase; G_2 = post-S interphase. Apart from the cortex cells just outside the quiescent centre (QC), which seem partly quiescent, the length of the cell cycle is similar throughout the meristem. (Barlow 1973)

	Hours				
	G_1	S	G_2	M	C
quiescent centre	135	16	13	6	170
root cap initials	−0.3	7.6	5.3	1.4	14.0
stele just outside QC	3.0	4.8	7.5	1.7	17.0
stele 1000 μm from QC	6.6	4.8	4.7	1.3	17.4
cortex just outside QC	21.8	10.2	6.4	4.3	42.8
cortex 1000 μm from QC	8.7	3.3	5.1	1.5	18.6

the lateral growth of the cells that are necessarily attached to it. Thus, in the root cap initials there are very few longitudinal divisions, so that the central part of the root cap tends to consist of a columella – columns of cells that are being displaced only towards the tip of the cap by the growth of the more recently formed root cap cells (Fig. 2.3). When a columella is present in a root cap, this therefore points to the presence of a QC. The converse, that roots lacking a root cap columella do not have a QC, is not necessarily true. Depending on the exact pattern of divisions in the root cap, a columella may not be obvious, even though there may be restricted lateral growth because of a QC. In roots with open meristems, in which there is more flexibility in the cell layers providing the particular set of initials operating at any one time, the QC may be more prone to fluctuations in size and activity.

The existence of a QC in roots with open and with closed meristems shows that the QC depends neither on one particular type of root organization nor on the presence of a thick wall between the epidermis and cap. The absence of a QC in young and small roots shows that it is not essential for root functioning. If it were the consequence only of the root being relatively broad, then this might suggest that it arises because of restricted access of essential substances to the centre of the meristem. Nutrients or other factors required for growth and cell division could perhaps be used up preferentially by the rapidly growing cells proximal to the QC. However, this would not explain why the root cap, on the far side of the QC, has some of the most rapidly dividing cells in the whole root.

When the root cap of maize is carefully removed so that the root is not otherwise damaged, the root grows faster (but ceases to respond to gravity) until a new cap is formed. Cell division in the QC is stimulated and slows down again only as the new cap regenerates. If the cap is cut longitudinally and only half is removed, cell division is stimulated only on that side of the QC next to the missing cap. This points to the root cap being both a restraint on the lateral growth of the root tip and a mechanical factor causing a QC to exist. An alternative explanation is that the root cap secretes an inhibitor of cell division. The loss of response to gravity when the cap is removed would also be consistent with the cap producing a growth regulator; so would the stimulation of division in the QC caused by cutting off only the distal part of the cap, leaving the cap initials and the base of the cap intact. Another suggestion has been that the QC is the source of some **growth substance**, which it supplies to the rest of the root and which is in supraoptimal concentration in the QC (its place of synthesis) and so inhibits cell division and growth. Cytokinin was a candidate for such a substance but immunocytochemical investigation has shown that cytokinin does not seem to be concentrated in the QC.

The cells in the QC divide much less frequently than other cells in the meristem and so are preserved as a reservoir of cells that have had less chance of alteration in their genomes during DNA replication. When the root tip is damaged, the QC is stimulated into division and can act as a source of cells to reconstitute the meristem. However, in nature, if a root is damaged it is perhaps more likely to be replaced by a new root originating as a lateral. It is also not clear whether there is a maximum number of cell divisions that take place in a cell before it gives up its meristematic function, as happens in the *Azolla* root. In the maize root, if a cell that remains in the QC divides about every 100 h then it would not divide more than about 60 times a season. Once it has been displaced out of the QC, a cell probably divides only four or five times more as it is displaced through the meristem. It seems possible that the promeristematic cells of many roots of higher plants could have a lifespan of about 60 generations, like *Azolla*. However, there are also roots, such as those of trees, that presumably can continue growing for many years without the root tip being replaced by a lateral meristem, and in which the cells may, therefore, have a much longer lifespan and divide many more times.

2.3.2   Shoot meristems[7]

Unlike the root, the shoot apical meristem has no cap and no QC, although in shoots the cells at the meristem summit grow and divide more slowly than their derivatives on the flanks of the meristem. This is true for apices with meristems or with apical cells (Table 2.2; Fig. 2.6). In *Helianthus* and *Nicotiana*, the cells of the peripheral parts of the vegetative shoot apices became labelled when [³H] thymidine (a DNA precursor) was

Table 2.2 Cell cycle in shoot apices. In vegetative shoot apices, the cell cycle is longer at the summit of the apex and shorter on the flanks where the leaves are initiated. (Lyndon 1976 with additions)

	Cell cycle (h)	
	Summit	Flanks
Oryza	86	11
Pisum	70	28
Trifolium	108	69
Rudbeckia	>40	30
Datura	76	36
Chrysanthemum	140	70
Solanum	117	74
Coleus	237	125
Sinapis	288	157
Isoetes (**pteridophyte**)	>53	36
Polypodium (fern)	144[a]	78
Hookeria (moss)	>105[a]	48
Polytrichum (moss)	360[a]	96

[a] Apical cell.

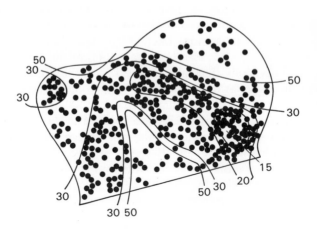

Figure 2.6 Rates of cell division in a pea shoot apex. In this median longitudinal section, the density of the points (colchicine metaphases) is proportional to the rate of cell division. The lines show the corresponding cell cycle lengths (h). The next leaf primordium is about to form on the right. (Lyndon 1973)

applied but the cells of the central zone did not, and so they do not divide and grow only slowly or not at all. In the converse experiment, the central zone cells of *Nicotiana* were labelled in excised apices, which were split and exposed to [³H] thymidine in culture. While the apices were regenerating the label became dissipated, during their growth, from all cells except those of the newly reconstituted central zone, which was therefore apparently not growing and dividing. Leaves are initiated only at the sides of the shoot meristem and not at the summit. Only when the flower is at the end of its initiation does the summit of the shoot apex become occupied by floral organs.

Control of division rates in the shoot apex may be by restriction of the growth rate because of the properties of the restraining structure, presumably the outer wall of the epidermis. There may also be an inhibitor of cell division produced at the tip of the meristem, lessening in effect with distance from the tip, although there is no evidence for this. The cells at the summit of the shoot apex often differ cytologically from the cells below and are therefore called the central zone. This does not correspond with (although is included in) the region of the slower division rate. The extent of the region of dividing cells is much greater in the shoot meristem than in the root meristem and encompasses probably most of the apical bud. Cell divisions are also frequent in the developing internodes and in the procambium. There is, therefore, not the same sharp cut-off of division or elongation as in the root. Relative growth rates in the shoot apex tend to be lower than in the root apex but, because of the much greater number of growing and dividing cells in the shoot apex, the absolute growth rate of the apical bud can be quite impressive.

2.3.3 Control of the cell cycle[8]

In the root, the cell cycle tends to be controlled by the entry of cells into DNA synthesis (S phase), as shown by the much greater variation in the length of G_1 than in the other phases of the cell cycle (see Table 2.1). The *Chrysanthemum* shoot apex is like the root – it is G_1 that is longer in the slower dividing cells at the apical summit. The shoot apex of *Pisum*, and also probably *Rudbeckia*, is different in that it is all the phases of the cell cycle (except mitosis itself) that are longer in the slower growing cells. There is, therefore, no hard and fast rule for the control of the cell cycle in the shoot.

In free-living cells, especially animal cells in culture, it has been proposed that the length of the cell cycle depends both on the length of time that the cells get stuck in G_1 and on the probability of the cells making the transition back into an actively cycling state. The length of the cell cycle would then depend primarily on the value for the probability of transition, and this would be reflected in the length of time spent by a cell in G_1. In *Saccharomyces cerevisae* (bakers' yeast), the cell cycle does not seem to be controlled in this way, but seems to depend on a minimum time in G_1 and the attainment of a minimal cell size before the next mitosis can take place. In plant root cells, G_1 can be non-existent in root

31

cap initials (Table 2.1), so there can be no minimum time in G_1 in these cells. Since plant cells are fastened to each other, the rate of elongation of adjacent cells must be the same, and so a sizer control would necessarily cause all cells to divide after a similar cycle length. The rate of elongation in the root meristem is indeed very similar for all cells (see Ch. 8, section 8.2). The transition probability model is most unlikely for the root, since we do not see occasional cells differing greatly in size from their neighbours, which is what would be expected.

The root meristem grows linearly and radially. Because of its geometry, the shoot meristem can grow in all directions and, therefore, it is theoretically possible for isolated cells to drop out of the cell cycle. They would not necessarily have to continue growing at their former rate but cells around them could continue faster growth except for those parts of their cell wall shared with the slower growing cell. In vegetative *Sinapis* shoot apices, there appear to be two subpopulations of cells. One has a cell cycle of about 86 h (33% of the cells) and the other (67% of the cells) has much longer cycles of up to 206 h. There is probably a complete spectrum of cycle lengths varying between the extreme values and differing mainly in the length of G_1. These subpopulations were shown by labelling to be completely mixed together within the meristem.

The cells in the root and shoot meristem are usually asynchronous, growing at similar rates but undergoing mitosis at different times. Some degree of synchrony can be found in small groups ('packets') of cells that are the product of the same mother cell. Synchrony can be induced in meristems by release from starvation but in rapidly growing cells it arises rarely, except during flowering.

2.3.4   Cell cycle changes in the flowering apex[9]

As shoot apices grow, mean cell size increases only very slowly so that essentially it remains constant. Growth rates are, therefore, directly proportional to rates of cell division and inversely proportional to the length of the cell cycle. On the transition to flowering, the growth rate of the apex usually increases, but only transiently (Table 2.3). As the flower is initiated and starts to form, the growth rate steadily decreases and the cell cycle lengthens (Table 2.4). This is despite the much more rapid initiation of primordia in the flower. This apparent paradox is possible because the floral organ primordia are smaller than the leaf primordia when they are initiated, so less growth is required to form one primordium. The apical dome usually enlarges on flowering but this is not necessarily because of the transient increase in growth rate but because relatively less of the apical dome is partitioned into each primordium at initiation and relatively more is retained as the apical dome (see Ch. 3, Fig. 3.8).

Table 2.3 Cell cycle in vegetative and flowering shoot apicés. The cell cycle usually shortens on flowering; *Epilobium* is the exception. (Lyndon 1976 with additions)

	Mean cell cycle (h)	
	Vegetative	Flowering
Triticum	41	22
Secale	50	31
Datura	36	26
Sinapis	157	25
Silene	20	10
Lupinus	48	34
Ranunculus	56	47
Epilobium	45	45

Table 2.4 Cell cycle in induction and flowering in *Silene*. The cell cycle lengthens while the flower is being initiated. (Francis and Lyndon 1985)

	Days from beginning of induction					
	0	1	2–7	8–9	10–13	14 . . .
photoperiod	SD	LD	LD	SD	SD	SD
cell cycle (h)	20	13	20	10	33	48
		induction		synchron- ization	flower initiation	flower development

In plants induced by photoperiod, the shoot apical cells become synchronized, as shown by successive waves of mitosis. In mustard (*Sinapis*), the period of evocation – between the arrival of the flowering stimulus at the apex and the appearance of flower buds – is occupied by a single more-or-less synchronous cell cycle of about 35 h (Fig. 2.7). This is accompanied, as would be expected in a normal cell cycle, by maxima and minima in the rates of RNA synthesis and by increases in the numbers of organelles before mitosis. Experimental treatments that prevent or delay this synchronous cell cycle also prevent or delay flowering, suggesting that the synchronous cycle is necessary for flowering.

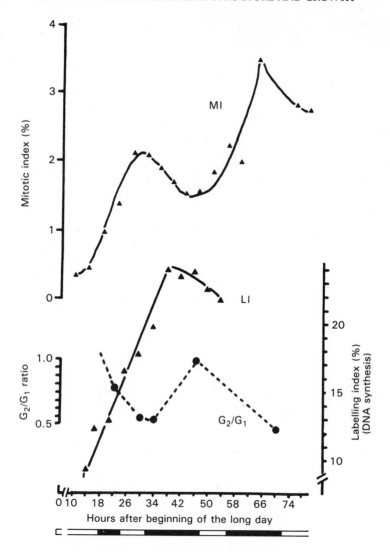

Figure 2.7 Synchrony during floral evocation in the *Sinapis* shoot meristem. In plants given one long day the apical cells become synchronized, as shown by the successive mitotic peaks preceded by peaks of cells in G_2 ($G_2 : G_1$ ratio) and with a peak of DNA synthesis in the middle of the cycle. (Francis & Lyndon 1985)

In *Silene*, this synchronization happens just before sepal initiation and can be prevented by placing the plants in darkness for 48 h after induction (Fig. 2.8); but the plants still flower, with a delay of 48 h. In *Silene*, therefore, synchronization is not essential for flowering. Synchronization has also been found in 3rd-order flower buds, which are not formed until several weeks after flower induction, and so here it is

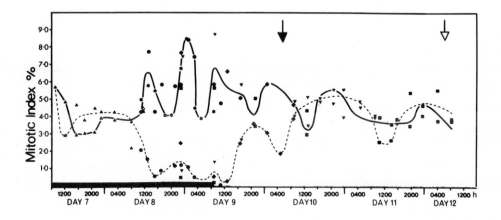

Figure 2.8 Suppression of synchrony but not of flowering in *Silene*. Plants induced to flower with seven long days were transferred to darkness for 48 h (----) before being restored to short days at 09.00 h on Day 9. The dark period inhibited apical growth and suppressed synchronization on Days 8 and 9 but the plants still flowered, but with a 48 h delay. Sepal formation in controls (—) (▼) and after 48 h darkness (▽). (Francis & Lyndon 1985)

thought to be a result of the apex becoming competent to form a flower. Synchronization seems to be an indicator of something else happening in the apical cells, which we have not yet discovered and which itself is essential for flowering.

In *Silene*, the cell cycle is also shortened at the very beginning of floral induction. Normally, seven long days are required to induce full flowering. However, as little as 5 min of far-red light at the very beginning of induction causes a shortening of just one cell cycle from 19 to 13 h. The effect of the light is so rapid that changes in the proportions of cells in different phases of the cell cycle can be measured cytologically within 1 h and imply a rapid increase in the number of cells in G_2. This could only come about if the cells are rushed through DNA synthesis. This is indeed what happens. The increase in the rate of replication fork movement from initiation points in the DNA molecules shows that the rate of DNA synthesis doubles within 30 min and probably as soon as the light pulse has been applied. Although the light pulse itself is not sufficient to cause floral induction, if it is replaced by a short (20 min) period of darkness, then flowering is 90% inhibited. Even though three long days are insufficient for induction, the first 5 min of the extended photoperiod nevertheless seems to be an essential part of induction. It is not known whether the increased rate of DNA synthesis and the shorter cell cycle during the first 13 h of induction are themselves essential for induction or whether, like the later synchronization, they are merely indicators of other essential, but as yet undetected, metabolic changes. It may be that changes in the apex during the first three long days make the *Silene* apex competent to respond to the next three long days by flowering (see Ch. 11). Some of the changes in the cell cycle on flowering can be mimicked by application of cytokinin, which causes cells to leave G_2 and divide. This causes a degree of synchrony, which is lost by the next division. Cytokinin also has no effect on DNA synthesis in the apex.

Meristems pose several problems of development:

(1) How is the integrity of the generating region (the promeristem or apical cell) maintained? This is not understood.
(2) How are the rates and planes of cell division controlled? (see Ch. 5).
(3) How is the shape of the meristem maintained and controlled? (See Ch. 4).
(4) How is the differentiation of cells controlled and the tissue pattern mapped out? (see Ch. 12).

2.4 SUMMARY

(1) Plants are constructed of iterated modules produced by the apical meristems. The shoot module is node + internode + leaf + axillary bud. The root module is root axis + lateral root.
(2) In lower plants the apical cell, which is often tetrahedral, is the ultimate source of all cells, which arise by subdivision and growth of the derivatives cut off in regular sequence from the faces of the apical cell. The sequence of divisions to form all the cells in the root has been followed in detail in the aquatic fern *Azolla*.
(3) Meristems of higher plants have several or many initial cells. In the root meristem the initials are within, or can be produced by, the quiescent centre – a group of cells with a very low rate of division. The highest rates of division are in those cells just outside the quiescent centre, that are eventually replaced by cells from the quiescent centre as it slowly grows and cells are displaced out of it. Isolated quiescent centres in culture can regenerate a whole root.

(4) The central summit cells of the shoot meristems of *Helianthus* and *Nicotiana* grow and divide very slowly, like a quiescent centre. In most shoot meristems, the cells at the summit grow and divide at about one-half or one-third of the rate of the cells on the flanks where the leaves are initiated.

(5) Clonal analysis shows that shoot apical initials can function for varying times but may produce both vegetative and reproductive structures – there is no germ line. Initials seem to function as such by virtue of their position rather than their lineage.

(6) The control of the rate of cell division in roots is mainly by control of entry of cells into the DNA synthesis (S) phase of the cell cycle. In *Chrysanthemum* shoot meristems control is also at the entry to S, but in other species the cell cycle in the shoot may be controlled by entry into S and into mitosis. Only mitosis itself is of relatively unvarying length. In plant meristems, cell sizes fall within a fairly narrow range and so the overall control of division rate may be by growth rate in conjunction with a sensor for cell size.

(7) On flowering, growth rates in the shoot meristem increase transiently but fall during flower development, even though smaller primordia are initiated faster. Photoperiod can alter the cell cycle in the shoot meristem and induce synchrony. As little as 5 min of light can increase the rate of DNA synthesis and shorten the cell cycle. Cytokinins, which induce a temporary partial synchrony but do not alter the length of the cell cycle in shoot meristems, mimic some of the changes that occur naturally on flowering.

FURTHER READING

Barlow, P. W. 1987. The cellular organization of roots and its response to the physical environment. In *Root development and function*, P. J. Gregory, J. V. Lake and D. A. Rose (eds), 1–26. SEB Seminar Series 30. Cambridge: Cambridge University Press.

Bryant, J. A. & D. Francis (eds) 1985. *The cell division cycle in plants*. SEB Seminar Series 26. Cambridge: Cambridge University Press.

Cutter, E. G. 1971. *Plant anatomy: experiment and interpretation. Part 2. Organs.* London: Edward Arnold.

Feldman, L. J. 1984. The development and dynamics of the root apical meristem. *American Journal of Botany* **71**, 1308–14.

John, P. C. L. (ed.) 1981. *The cell cycle.* SEB Seminar Series 10. Cambridge: Cambridge University Press.

Lyndon, R. F. 1976. The shoot apex. In *Cell division in higher plants*, M. M. Yeoman (ed.), 285–314. London: Academic Press.

NOTES

1 Gunning *et al* (1978), Gunning (1982). (*Azolla*)
2 Cutter, (1954), Gifford (1983). (Apical cells)
3 Barlow (1987), Clowes (1981), Feldman & Torrey (1976). (Structure and functioning of root meristem initial cells).
4 Shabde & Murashige (1977). (Culture of shoot meristems)
5 Coe & Neuffer (1978), Jegla & Sussex (1987), Poethig (1987) Tilney-Bassett (1986). (Clonal analysis and chimeras)
6 Barlow (1973), Barlow & Hines (1982), Clowes (1972). (Cell division in the root and root cap removal)
7 Lyndon (1973; 1976). (Cell division in shoot apices)
8 Gonthier *et al.* (1985). (*Sinapis*)
9 Francis & Lyndon (1985), Gonthier *et al.* (1987), Lyndon & Francis (1984), Ormrod & Francis (1986). (Cell division in flowering shoot meristem)

CHAPTER THREE

Meristem functioning:
formation of branches, leaves,
and floral organs

3.1 FORMATION OF BRANCHES

Branching in plants with apical cells is usually by division of the apical cell giving two meristems and so producing dichotomous, or false dichotomous, branching. Moss leaves do not have axillary buds. Axillary meristems are present at the base of the leaves in many ferns but they do not always grow out to form stem or rhizome branches, and sometimes they may form new fronds. In higher plants, the shoot branches by the outgrowth of the axillary buds to form new side shoots. The branched root system is formed by lateral roots, which themselves produce laterals, and so on.

3.1.1 Lateral roots[1]

Lateral roots in higher plants originate in the **pericycle**, usually some distance back (1–2 cm) from the root tip. The first sign is increased basophilia of the pericycle cells, indicative of protein and RNA synthesis, followed by periclinal divisions and subsequently anticlinal divisions, to give a new meristem. Each lateral originates from a group of about 20–30 pericyclic cells. Lateral roots are initiated behind the root tip from late-maturing pericyclic cells that remain capable of division and are probably still progressing through the cell cycle. These pericyclic cells, and so the ranks of laterals that are formed from them, are nearly always opposite the **xylem poles**, perhaps because substances diffusing from the xylem (auxins?) delay pericyclic maturation. In addition to these laterals that form naturally, extra (adventive) laterals can be stimulated to form by supplying auxin, which in radish roots causes pericyclic cells in G_2 to go on to divide within 2 h and resume cycling (Fig. 3.1). Lateral roots can

39

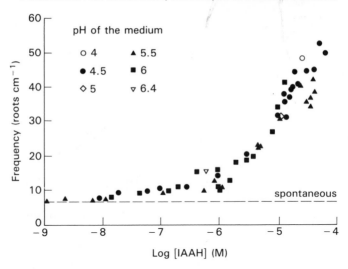

Figure 3.1 Auxin-induced lateral root initiation. About six laterals per centimetre form spontaneously. Increasing auxin concentration in the culture medium causes a two-phase stimulation of lateral formation as a function of the log concentration of undissociated IAA [IAAH], the form in which it probably passes across the endodermis to the pericycle. (Blakely *et al*. 1988)

be made to initiate in continuous vertical rows, each lateral abutting on its neighbour, by increasing the concentration of auxin in the culture medium. The emergence of lateral roots through the cortex of the parent root may be facilitated by breakdown of the cortical cells. Lateral roots of *Vicia* require for growth the carbohydrates in solution in a cavity that forms in the cortex ahead of the young lateral. The presence of this cavity implies that the parent root cells are degraded enzymically ahead of the growing lateral, which does not therefore simply push its way out.

Where the lateral roots are formed much nearer the root tip, as in some water plants, endodermal and cortical cells can also be involved in the formation of the lateral meristem. These cells seem to become less able to redivide and grow as they mature, so the further back laterals are initiated the more exclusively they are formed from only those pericyclic cells that have differentiated least and so retained most of their meristematic potentiality.

In the ferns, the lateral roots form not from the pericycle but from the next cell layer out, the **endodermis**. Unfortunately, *Azolla* (see Ch. 2, section 2.2.1) does not form lateral roots. In other ferns, the position of lateral root initiation is predictable from the cell lineage at specific positions in the root. In *Marsilea*, laterals arise from large endodermal cells opposite the protoxylem, which have become large by omitting one cell division while continuing to grow. They subsequently divide to form derivatives that include a new apical cell, which then divides in a right- or

40

left-handed fashion. The handedness adopted depends (in a way that is not understood) on the position and orientation of the stem that bears the parent root. The large endodermal cells alternate with small cells that continue division to extend the endodermis.

3.1.2 Lateral shoots

In higher plants, lateral shoots develop from axillary meristems formed in the axil of each newly initiated leaf primordium. The axillary bud develops from superficial cells and is therefore exogenous in origin, whereas the roots are formed from internal tissues and are endogenous. Adventive shoot buds, especially in callus, can sometimes arise endogenously too. The axillary bud develops a typical apical dome and begins to form leaf primordia, the positions of which are usually in characteristic positions and orientation relative to the parent shoot (see section 3.2.2). Often, the control of subsequent bud growth is by the main shoot apex, the apical dominance of which becomes less as it grows on upward and becomes more distant from the bud.

3:2   LEAF INITIATION[2]

The shoot meristem differs from the root in that it forms organs on its sides. These develop into leaves or floral parts. In the root, there is no counterpart to the leaves; the lateral roots are analogous to the shoot axillary buds. The shape of the shoot meristem changes as each leaf is initiated on the flanks of the meristem. This implies that the apparently rigid controls of form, which in the root maintain its cylindrical shape, can be relaxed temporarily on the sides of the shoot meristem. When the shoot meristem ultimately becomes a flower meristem, organs form on its summit so that it is all used up to form floral parts and its growth comes to an end.

The node and internode are initiated at the same time as the leaf primordium. In *Sambucus*, the internode and node are apparently each initiated as a single layer of cells at the shoot apex (Fig. 3.2). In *Silene*, there seem to be four layers of cells associated with each leaf, two of these layers giving rise to the nodal cells and the other two to internodal cells. Other plants may be similar. If so, the structure of the shoot apical meristem is more ordered than appears from histological preparations. Support for these ideas comes from clonal analysis (see Ch. 2, Box 2.2.), which shows that files of cells derived from common initials at the apex always end at the top or the bottom of an internode, rarely in the middle. This is consistent with the whole internode being formed from a single cell layer in the apex.

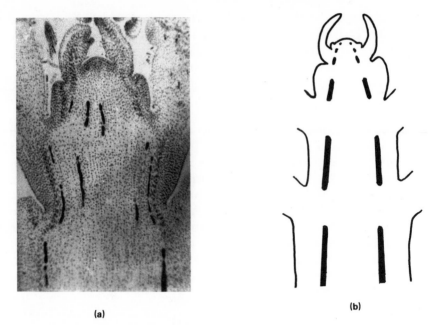

(a)

(b)

Figure 3.2 Origin of internodes. In *Sambucus racemosa*, coenocytic tannin-filled cells extend the length of each internode. (b) These cells, and therefore the internodes, seem to originate from alternate cell layers in the shoot apex. The intervening single cell layers give rise to the nodes. (Zobel 1985: photograph kindly supplied by Dr A. Zobel)

3.2.1 Mechanism of leaf formation[3]

Changes in shape and maintenance of shape can be achieved in several ways (see Ch. 4, Box 4.1). In the formation of leaf primordia on the shoot apex, the existence of a 'growth centre' of increased growth rate at the site of leaf initiation is often assumed. However, only direct measurements can show whether local changes in growth rates or growth directions are the main factors in leaf initiation.

In the shoot apices of pea (*Pisum*) and clover (*Trifolium*), there is little difference in the growth rate on the opposite sides of the apex, only one of which is about to make the next leaf (see Ch. 2, Fig. 2.6). The cell division rate is also high in the axial part of the apex, which does not change much in size or shape when a leaf primordium begins to be formed. These facts suggested that the main cause of the formation of a leaf primordium was a change in the direction, rather than the rate, of growth. This was confirmed by finding that the plane of cell division, which is normal to the direction of growth, changed in the pea apex about 16 h before a leaf primordium could first be seen as a slight bulge. At the leaf site, periclinal divisions appeared where there were none before (Fig. 3.3).

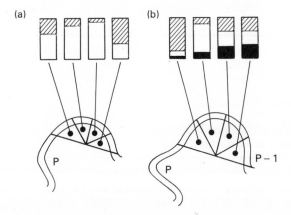

Figure 3.3 Periclinal divisions and leaf initiation in pea. (a) Just after a new leaf primordium (P) has begun to grow out, all divisions in the apical dome are anticlinal in the plane of the section (□) or perpendicular to it (▨). (b) About 16 h before the next primordium is formed at P-1, periclinal divisions (■) appear. The divisions in the three planes are in the ratio 1:1:1, indicating no preferred division plane. The restriction on periclinal division in (a) has been lifted in (b). (Lyndon 1983)

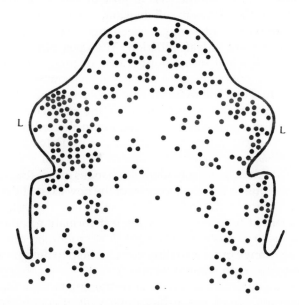

Figure 3.4 Rates of cell division in the *Silene* shoot apex. The rate of division is proportional to the points (colchicine–metaphases). Division is fastest in the young leaf primordia (L), unlike *Pisum* (cf. Fig. 2.6).

Table 3.1 Rates of cell division during leaf initiation in the pea. Cells at the position at which the next primordium will form (I_1) divide a little faster only when the primordium (P) is beginning to bulge out. The rate then falls as the primordium grows. (After Lyndon 1983)

	Rate of cell division (% cells h^{-1})	Mean cell cycle (h)
I_1: first part of plastochron	2.5	27.7
second part of plastochron	2.5	27.7
P: first part of plastochron	2.9	23.9
second part of plastochron	2.0	34.7

In the apex of *Silene*, on the other hand, there is no apparent change in the planes of cell division when a leaf pair is formed but the rate of growth and division at the leaf sites is increased (Fig. 3.4). *Silene* and *Pisum*, therefore, seem to exemplify the different extremes in the way in which the growth direction is changed when a leaf is formed. In *Pisum*, there is mainly a change in division planes without much change in growth rate, except transiently as the primordium begins to form (Table 3.1). In *Silene*, there is an increase in growth and division rate without much change in the planes of cell division.

Whether growth or division planes change, the apex must be able to bulge out at the site of leaf initiation. The apex can be regarded as a system in which a leaf forms not because of the imposition of a new growth direction but because the restriction on growth in all directions is released locally. For a leaf to form at all, the surface must be deformable. This presumably requires an increase in the plasticity of the surface, irrespective of the microfibrillar orientation in the cell walls (see Ch. 4, section 4.4.1) or the rates of growth of the underlying cells. Without periclinal divisions in these cells, bulging at the leaf site would not be possible unless there were considerable changes in cell shape so that they became very elongated in the direction of the new axis. The occurrence of periclinal divisions and an adequate growth rate at the potential leaf site may be regarded as prerequisites for leaf formation, and changes in surface plasticity as its cause. Changes in microfibrillar orientation may facilitate the change in growth direction and help to determine the shape of the emerging organ when it forms.

The obvious candidate for an outer restraining layer in the shoot apex is the epidermis, which has a thick outer wall. It is also the tissue that responds to auxin in elongating stems (see Ch. 8, Box 8.2), but whether the growth of the epidermis at leaf initiation sites is controlled by endogenous auxin is not known. Evidence from a chimera points to a possible role of the epidermis in controlling primordium initiation. In the

Camellia chimera 'Daisy Eagleson', the inner cells are from *C. japonica*, which has only sepals and petals but does not form stamens or carpels, and the epidermis is from *C. sasanqua*, which forms all floral organs. The chimera (in which only the epidermis is from *C. sasanqua*) forms all types of floral organs, so clearly the epidermis is responsible in some way for the formation of stamens and carpels in the chimera. How the epidermis may be involved in primordium formation is considered in Chapter 4 (section 4.4).

3.2.2 Positioning of leaf primordia[4]

Whatever the mechanism of leaf primordium initiation and formation, it must account for not only how a primordium is formed but also when and where on the shoot apex. Leaf primordia are initiated only at the sides of the apex. If leaf initiation occurred simultaneously all round the flanks of the apex, then a collar would be formed. This can be achieved by applying TIBA, an inhibitor of auxin transport, to the apex and implicates auxin distribution as a probable controlling factor in the siting of primordia. TIBA would presumably raise the concentration of auxin in the apex by preventing it from being transported away. This would be consistent with an increased local concentration of auxin being necessary for a primordium to form. For normal primordium formation, this increased auxin concentration would have to be restricted to only a part of the apex. Another inhibitor of auxin transport, NPA, when spotted onto a very young primordium of *Epilobium* caused it to occupy a larger tangential area. PCIB, an inhibitor of auxin action, caused a primordium to be smaller in extent. Both treatments caused an alteration in subsequent **phyllotaxis** (leaf arrangement), consistent with the size of one primordium affecting the position of the next, and therefore all subsequent, primordia. All these treatments are consistent with auxin being involved in determining primordium size by controlling the area over which the apical surface can bulge outwards. When 2,4-D, an auxin analogue, is applied to apices it can cause the formation of increased numbers of leaves, again suggesting a role for auxin in promoting primordium formation. TIBA can also alter phyllotaxis by making the apex narrower and so effectively increasing the primordium : apex area ratio. The size of a primordium when it is initiated is important because it determines how many primordia can be accommodated on the apical surface and, therefore, is a prime determinant of the arrangement of primordia and hence the arrangement of leaves.

The arrangement of leaves at the apex, or phyllotaxis, is such that it gives rise to consistent and predictable patterns. The most striking are those of distichous leaves: in two ranks where successive leaves are exactly 180° away from their neighbours, as in the pea; and the opposite

(a)　　　　　　　　　　　　　　　　　　(b)

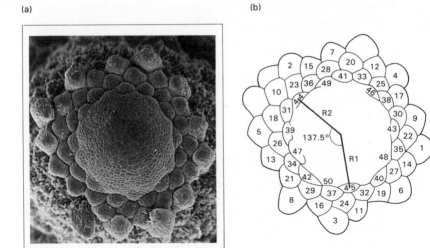

Figure 3.5 Spiral phyllotaxis. (a) Scanning electron micrograph looking down on a shoot apex of *Pinus contorta*. (b) The primordia are numbered in the order in which they were initiated (oldest is 1). The divergence angle is approximately 137°. The plastochron ratio is R2/R1. (After Cannell 1976; photograph kindly supplied by Dr M. Cannell)

and decussate arrangements, as in Labiates such as the deadnettle (*Lamium*), in which each leaf of an opposite pair is exactly 180° distant from its twin and each pair is exactly 90° displaced from the next pair up or down the stem. These exact placings of leaves have, paradoxically, been more difficult to understand than the more complicated arrangements.

It has long been a source of fascination that in many plants the leaves are arranged at the apex in a spiral (more properly a helix) in which each successive leaf is displaced from its previous neighbour by a divergence angle of about 137° (Fig. 3.5). The significance of 137° is that it is essentially the limiting angle approached in the expression

$$360 \left(1 - \frac{n-1}{n} \right)$$

where n and $n-1$ are the larger and smaller of successive numbers in the Fibonacci series 1,1,2,3,5,8,13,21,34... – an infinite series that approaches the limit

$$\frac{\sqrt{5}-1}{2}$$

The expression is, therefore, more precisely written as

$$360 \left(1 - \frac{\sqrt{5}-1}{2} \right)$$

46

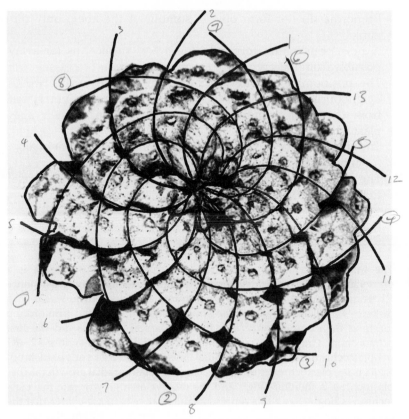

Figure 3.6 Phyllotaxis on a pine cone. These are scales rather than leaves but clearly show the two sets of parastichies, 8 in one direction and 13 in the other.

The consequence is that regular helical leaf arrangements can be described mathematically and the formation of successive leaves at the divergence angle of 137.5° gives rise to patterns having Fibonacci properties. Such pattern can be seen in pine cones in which the winding lines of scales can be traced in two opposite directions, the number of lines (parastichies) in one direction corresponding to a number in the Fibonacci series and the number in the opposite direction corresponding to an adjacent number in the same series (Fig. 3.6). The problem is to account for the divergence angle of 137.5°, or something very close to it, being adopted by many plants.

F. J. Richards showed that a spiral of leaves showing parastichies but no orthostichies (radii from the apical centre connecting any pair of leaves, however distant) would be produced if an apex initiating leaves did so according to a set of rules that could be formulated. There were:

(1) Primordia do not form on the summit of the apex, only on the flanks.

(2) A new primordium forms on the apical surface, as far away as possible from the previous primordium.

(3) The positioning of a new primordium relative to the two older neighbouring primordia is such that the new primordium is distant from the older primordia in the inverse ratio of their ages in plastochrons.

(4) The apical surface expands at a rate that increases exponentially as a function of the radius (Box 3.1).

(5) The area occupied by a primordium at initiation is constant relative to the area of the apical surface.

Box 3.1 Measurement of growth using the plastochron ratio[5]

The interval between the initiation of successive leaf primordia is a plastochron, a measure of developmental time. After initiation at a distance $R1$ from the apical centre, the primordium is displaced outwards to a distance $R2$ by the growth of the shoot meristem. The next primordium forms at the same distance, $R1$, from the apical centre as did the first primordium. The radial expansion of the apical dome is therefore $R2-R1$ per plastochron and the relative radial expansion is $R2/R1$ per plastochron. $R2/R1$ is the plastochron ratio r (Fig. 3.5). The relative radial growth rate per plastochron is therefore $\log_e r$ and the relative area growth rate per plastochron is $2\log_e r$. Since this represents the relative increase in area that the apex has had to make in order to initiate a new primordium, this can be taken as a measure of the area of a primordium at initiation relative to the area of the apical dome. If growth rate in volume is assumed to be $3\log_e r$, then the mean doubling time in plastochrons for the apical tissues is $\log_e 2/3\log_e r$. If the plastochron is known, by measurement of the rate at which leaf primordia are formed, then $(\log_e 2/3\log_e r) \times$ plastochron in h gives the mean cell cycle length (h) for the cells of the apex. Measurements of the plastochron ratio have allowed calculation of the cell doubling time in the oil palm (*Elaeis*) shoot apex as 1150 h, which is surprisingly long considering that this is a plant of the humid tropics. The plastochron ratio has also been used to show that the apical growth rates of *Pinus contorta* trees of different provenances were very similar and that the faster growing trees were the result of needle production over a longer period and not because of faster growth.

Rules 1–4 are consistent with observation and measurement. Rule 5 is an assumption. The 2nd and 3rd rules would be consistent with the young primordia producing an inhibitor of primordium initiation that decreases in effectiveness with increasing age of the primordium, new primordia being sites of inhibitor production. Rule 5 implies a minimum area of apical surface that needs to be generated by the continued growth of the apex before a new primordium can form (i.e. competence to form a primordium depends on a minimum available surface area). An attraction of this hypothesis, which is essentially a 'field' theory of phyllotaxis, is that the positioning of new primordia does not have to be particularly precise because the system is self-adjusting owing to the influence of the older primordia. Indeed, slight imprecision in positioning would allow spiral Fibonacci phyllotaxis to become established on an apex where the first two primordia are opposite, as the cotyledons are in **dicotyledonous** seedlings, but a spiral leaf arrangement is soon formed.

An opposite, modelling approach has also been used to account for Fibonacci phyllotaxis. Starting from the assumptions that a new primordium is the site of production of an inhibitor of primordium initiation, that the inhibitor decays in effectiveness with time, and that new sites of inhibitor production form at minima in the inhibitor field, Thornley has shown that such a system would generate a succession of inhibitor-producing sites separated from each other by a divergence angle of 137.5°. This model shows that if new primordia were sites of inhibitor production, and these were formed at minima of inhibitor concentration, then they would position themselves according to Rules 2 and 3 and Fibonacci phyllotaxis would result.

These hypotheses are concerned with explaining a given leaf arrangement and, therefore, assume a constant size for the primordia at initiation (Rule 5). This is convenient, as it avoids the further complication of a changing phyllotaxis. If primordium size changed, then this in itself would alter phyllotaxis, as happens on flowering (see section 3.3) or in plants treated with auxin antagonists (see above). Note that an increase in size of the whole system will have no effect on leaf arrangement. The crucial factor in determining phyllotaxis is the size of the primordia relative to the size of the apical surface, i.e. how the apex is partitioned at primordial initiation.

Some other theories of primordium positioning assume that a new primordium occupies the 'first available space' on the apical surface. These are essentially packing theories and depend on how the availability of space is determined. If this is ultimately at the cellular and biochemical level, as it surely must be when the primordium becomes determined (see Ch. 10, section 10.8.2), before it becomes a morphologically distinct entity, then they become difficult to distinguish from field theories. The first available space theory implies that new primordial sites are physically distinct from each other and should not, therefore, be able to overlap, a necessity not prescribed by field theories. The observation that, occasionally, leaves can be formed fused together and so new leaf sites can indeed overlap each other seems to preclude the first available space theory, except insofar as it implies that there is some minimum area of apical surface necessary for a primordium to form at all. The field theories of primordium initiation are consistent with the primordia being sites of production of an inhibitor that prevents the initiation of new primordia. Note that this hypothesis implies that new sources of inhibitor production would have to arise at concentration minima in the inhibitor field, i.e. low concentrations of a substance (the inhibitor) would induce its own synthesis. This parallels what is thought to happen in habituation in callus (see Ch. 10, Box 10.1). An attraction of field theories is that the time and position at which a new primordium is formed can be modelled in terms of the rates of processes. A more recent theory is that the biophysical microstructure of the epidermal cell walls is the determinant factor in primordium initiation. In Chapter 4, we shall see how this may be complementary to field theories in explaining how the epidermis grows faster, as it must, where it bulges to form young leaves or other primordia.

The procambium has been thought to be another possible influence on the control of leaf initiation. Since procambium seems to differentiate **acropetally** in shoots, its development appears to be controlled from below. The procambial strands also go into the young leaf primordia. Larson has shown that, in cottonwood trees (*Populus*), procambium differentiation can be detected below the sites of future leaves up to 14 plastochrons before the primordia themselves become visible and, therefore, long before they are initiated. The position of the procambium, therefore, predicts the sites of future primordia. However, it could also be that because of the complexity of the vascular network in these trees, the development of procambium between existing strands is almost bound to be beneath the site of one leaf or another. Other evidence, from tissue culture, is that new shoot meristems can arise in callus and begin to form leaf primordia in the phyllotactic arrangement characteristic of the species from which the callus is derived. In such cases of *de novo* formation of apices with primordia, there is no pre-existing procambium that could possibly determine primordial sites. Also, in experiments in which incisions were made in *Lupinus* apices it was possible to alter the position of the next-but-one primordium to be formed by a transverse cut that severed it from all possible influence of procambium below it. What seems most likely is that procambium differentiaties in strands under the influence of primordia (see Ch. 7, section 7.6) and that procambium may also form in relation to existing vascular tissues some distance below the primordia. The final course of the procambium presumably depends on influences from both the developing primordium and the existing vascular system below it.

3.3   CHANGES IN PRIMORDIAL INITIATION ON FLOWERING[6]

The organization at the apex typically changes on flowering (Table 3.2). The precocious outgrowth of axillary buds shows that there is a release of apical dominance on flowering. If this is normally imposed by auxin, this would suggest a possible reduction in auxin availability at the apex when flowering begins. This would also be consistent with the reduction in size of primordia at initiation in the flower if auxin is involved in determining primordial size (see section 3.2.2). At their initiation, the sepals are smaller than the leaves, and the petals and stamens are smaller again (Table 3.3). The reduction in primordial size goes hand in hand with a changed primordial arrangement. Smaller primordia relative to the apical dome surface area should lead to higher order phyllotaxis, i.e. to the Fibonacci divergence angle (137.5°) and to a more complex spiral arrangement of parastichies. In flowers, this point is not in fact reached. In **monocotyledons**, the floral parts are usually in threes, i.e. a divergence angle of 120° between successive primordia; in dicotyledons, the floral parts are usually in fours or fives. Because the floral organs are often formed sequentially (all members of a floral whorl not necessarily being initiated simultaneously), the true divergence angle in a pentamerous (5-membered) whorl is usually 144°, so that the final angle between neighbouring members is 72° (360/5) (Fig. 3.7). This is the divergence angle that would be achieved on the inhibitor model if the effects of the inhibitor lasted only two plastochrons after a primordium was initiated.

Although an increase in growth rate of the apex is characteristic of the transition from making leaves to making floral organs, it is not this that alters the arrangement of the primordia (Box 3.2). The critical change is that the apical dome increases in surface area relative to the primordia at initiation. This could be brought about either by an increase in surface area of the apical dome by more rapid growth than on the flanks where the primordia are initiated (this, in fact, tends to happen on flowering when the summit of the apex is activated to grow more rapidly), or by the primordia becoming smaller at initiation relative to the apical surface whether or not the apex grows faster or bigger (Fig. 3.8). The importance of the partitioning of the apex rather than the growth rates is shown by those few apices (e.g. *Humulus* and *Perilla*) that decrease in size just before flowering, but in which the characteristic change in primordial arrangement still takes place. The control of primordium size at initiation is, therefore, a crucial process in apical functioning but one of which we know little. At the same time as primordial size is reduced, the amount of tissue to form the stem associated with each primordium is also reduced so that internodes do not form in the flower. This may also be because the cells that normally extend to form the vegetative internodes are not

Table 3.2 Changes in organization of the shoot apex on flowering.

In the shoot
1 Precocious axillary bud growth, which leads to change in branching pattern
2 Change in leaf form (bract formation)
3 Change in phyllotaxis to give more complex system

At the shoot apical meristem
1 Growth rate increases temporarily
2 Apex usually enlarges
3 Rate of initiation of primordia increases
4 Cellular changes associated with evocation: synthesis of new RNA and proteins and increase in mitochondria and rate of respiration

In the flower
1 Primordia are smaller at initiation (relative to the apical dome)
2 Changes in divergence angle between primordia so that arrangement of the primordia (phyllotaxis) changes to give whorls
3 Internodes are suppressed
4 Loss of axillary structures
5 Precise sequence of floral whorls: sequence of states
6 Primordia differentiate along new developmental pathways
7 Primordia occupy the summit of the meristem
8 Form of flower constant: interactions of primordia of different whorls
9 Meiosis
10 Cessation of growth until fertilization

Table 3.3 Reduction in size of primordia at initiation during flower formation. The area (relative to apical dome area = 1) of the floral primordia is much less than the leaves. (Lyndon & Cunninghame 1986)

	Primordium area
Ranunculus	
leaves	0.71
sepals	0.23
stamens	0.03
Impatiens	
leaves	0.41
petals	0.26
Silene	
leaves	0.33
sepals	0.07
petals	0.04
stamens	0.08

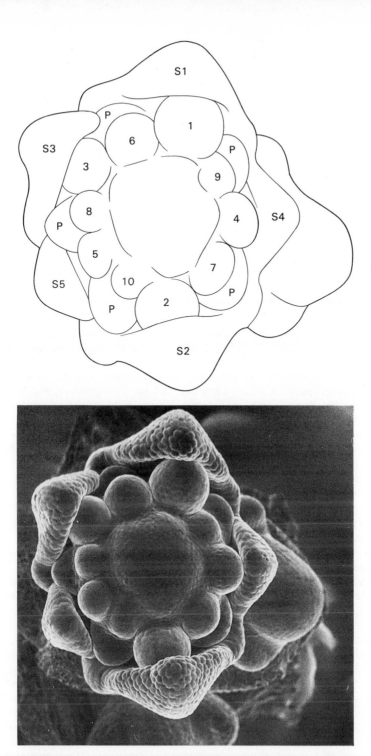

Figure 3.7 Primordium arrangement in a flower. In this *Silene* flower, only the carpels have not yet been initiated. The divergence angle between successive sepals (S1, S2, etc.) and stamens (1,2, etc., 1 being the oldest) varies considerably from 156° (S1–S2) to 104° (stamens 5–6), and reflects the change from the twofold symmetry of the vegetative plant with opposite leaves to the fivefold symmetry of the flower, which develops as the flower develops. P = petal.

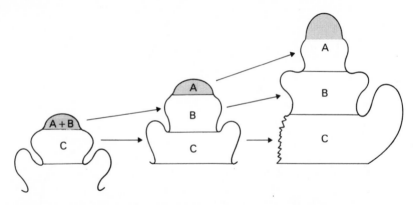

Figure 3.8 Enlargement of the apical dome. With a constant relative growth rate of the apex but different partitioning between primordia and apical dome, the result is enlargement of the apical dome. Even with a reduction in growth rate, flower organ primordia can be initiated faster than leaf primordia because they are smaller.

Box 3.2 Growth and flowering at the *Chrysanthemum* shoot apex[7]

When *Chrysanthemum* flowers, the apex enlarges. There is a 400-fold increase in apex area within a few days to form the capitulum, which bears the many florets. The apex forms leaves, then **bracts**, then florets. Cockshull and his colleagues have shown that the first bract always forms when the apex reaches a diameter of 0.26 mm, irrespective of the environmental conditions used to achieve flowering or the growth rate of the plant.

This has been modelled by assuming that (as observed) the first bract forms when the apical dome has reached a critical size (see Fig. 3.8) (although why this should happen is left unanswered), and that bract and floret primordia compete with the apical dome for assimilates whereas the leaf primordia do not. As a result of introducing this competition factor, the apex is eventually used up in the formation of the floret primordia, and changes in primordium size at initiation, apex growth rate, and apical size, similar to those also found in other species, can be accounted for. This model provides for a change in growth pattern after floral transition and is able to predict accurately the number of florets that will be formed in different environments.

A second model uses the equations of catastrophe theory to simulate the switch from vegetative to floral growth. According to this model, the inputs (growth rate, plastochron, etc.) remain constant but the system reaches an unstable point and shifts to a new stable state, i.e. flowering. The changes predicted by this model at transition (Fig. 3.9) include a decrease in primordium size, a shortening of the plastochron, and an immediate increase in apical growth rate, all of which are characteristic changes in apices on flowering (Table 3.2). This model is of interest because it implies that a sustained pattern of growth can nevertheless bring about an abrupt change of state once the system reaches a critical point. This is reminiscent of the observation that usually plants will eventually flower in any

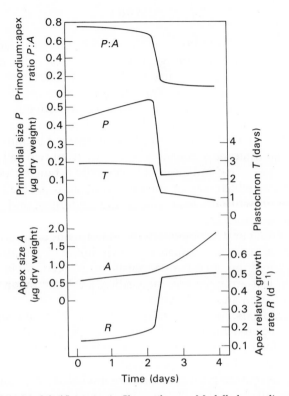

Figure 3.9 Model of flowering in *Chrysanthemum*. Modelled according to catastrophe theory, constant inputs can result in a sudden switch to a new stable state with changes in primordial size and initiation rate, and apical size and growth rate, of the sorts commonly observed on flowering. (Thornley & Cockshull 1980)

conditions in which they can grow if they are left long enough. A basic assumption of both of these models is that the newly initiated primordia are sources of a morphogen that inhibits the initiation of new primordia until its concentration has fallen to a threshold value. The plausibility of these models for *Chrysanthemum* lends some credence to the existence of such morphogens, but until they can be demonstrated they remain purely hypothetical.

formed in the developing flower. This becomes a problem of determination of specific cells or cell layers in the apical dome (see section 3.2). How this is controlled is as yet unknown.

3.4 SUMMARY

(1) Lateral roots are initiated from pericyclic cells that are probably still cycling and are late-maturing. The initiation of extra adventive laterals can be stimulated by auxin, which in radish causes pericyclic cells in G_2 to resume cycling. The lateral may enzymatically dissolve its way out through the parent cortex. Lateral shoots form from axillary buds that can be inhibited by apical dominance.

(2) The initiation of a leaf primordium is accompanied by initiation of a node and an internode, probably originally from two cell layers in the shoot apex. Leaf initiation is necessarily a bulging out of the apical surface, which may be the key event, accompanied by some increase in growth rate, especially in the epidermis, and periclinal cell divisions, which are probably a prerequisite rather than a cause.

(3) The positioning and arrangement of primordia depend on the area on the apical surface occupied by newly initiated primordia. This can be altered by auxin antagonists with predictable ensuing changes in leaf arrangement, suggesting a role for auxin in primordium initiation and phyllotaxis.

(4) Spiral phyllotaxis, with a divergence angle between successive primordia approaching the Fibonacci angle of 137.5°, can be accounted for by primordial initiation following rules that can be derived from, and are consistent with, observation of apical functioning. Primordium positioning can be modelled by assuming that the newly initiated primordia are sites of production of an inhibitor of initiation of further primordia, that they arise at concentration minima in the inhibitor field, and that the effectiveness of the inhibitor declines with primordium age. Whether such inhibitors really exist is unknown.

(5) There are many theories of phyllotaxis, many of them essentially packing theories ultimately requiring explanation in cellular terms. The procambium seems to direct the placing of primordia, but this may be a result of both procambium and primordia arising at predictable positions in an ordered system. Normal phyllotaxis in isolated, cultured shoot apices argues for autonomy of the apex.

(6) On flowering, the primordial arrangement changes to the characteristic tri-, tetra- and pentamerous patterns of flowers. Primordia in the flower are smaller at initiation than in the leaves, and this is probably the key factor in the change in primordial arrangement. It focuses attention on the need to understand the regulation of primordial size and the partitioning of the apex between primordium and apical dome each time a primordium is formed.

FURTHER READING

Barlow, P. W. 1984. Positional controls in root development. In *Positional controls in plant development*, P. W. Barlow & D. J. Carr (eds). 281–318. Cambridge: Cambridge University Press.

Cutter, E. G. 1971. *Plant anatomy: experiment and interpretation. Part 2. Organs.* London: Edward Arnold.

Hillman, J. R. 1984. Apical dominance. In *Advanced plant physiology*, M. B. Wilkins (ed.), 127–48. London: Pitman.

Lyndon R. F. & M. E. Cunninghame 1986. Control of shoot apical development via cell division. In *Plasticity in plants*, D. H. Jennings & A. J. Trewavas (eds). Symposia of the Society for Experimental Biology **40**, 233–255.

Lyndon R. F. 1983. The mechanism of leaf initiation. In *The growth and functioning of leaves*, J. E. Dale & F. L. Milthorpe (eds), 3–24. Cambridge: Cambridge University Press.

Richards, F. J. & W.W. Schwabe 1969. Phyllotaxis: a problem of growth and form. In *Plant physiology: a treatise*, Vol. VA, F. C. Steward (ed)., 79–116. New York: Academic Press.

Schwabe, W. W. 1984. Phyllotaxis. In *Positional controls in plant development*, P. W. Barlow & D. J. Carr (eds), 403–40. Cambridge: Cambridge University Press.

NOTES

1 Blakely *et al*. (1982), Blakely *et al*. (1988), Macleod & Francis (1976). (Lateral root formation)
2 Zobel (1989), Lyndon (1987). (Formation of nodes and internodes)
3 Stewart *et al*. (1972). (*Camellia* chimera)
4 Larson (1983), Meicenheimer (1981), Thornley (1975), Tran Thanh Van (1981), Schwabe (1971). (Positioning of leaf primordia)
5 Cannell (1976). (*Pinus contorta*)
6 Lyndon (1978), Lyndon &Battey (1985). (Apical changes on flowering)
7 Charles-Edwards *et al*. (1979), Thornley & Cockshull (1980). (Models of flowering)

PART III

Control of shape
and directions of growth:
the cellular basis of form

CHAPTER FOUR

Shape, growth directions, and surface structure

4.1 MAINTENANCE OF SHAPE[1]

The formation of leaves, the outgrowth of axillary buds, and the formation of lateral roots all involve the formation of a new growth axis at right angles, or normal, to the previous axis. We need to understand: (1) how the growth direction of the original axis is maintained; and (2) how a new axis is formed, i.e. how is the new direction of growth initiated and maintained? We also need to know the relative contributions of changes in the rates and orientations of growth to the growth process (Box 4.1).

The dome-like shape of the typical shoot apex and the tunica–corpus

Box 4.1 The basis of changes in shape

Changes in the shape of organs and plant parts can occur in more than one way. For instance, a cylinder growing to a trumpet shape could do so in basically two ways. In the first, growth is isotropic (equal in all directions) but there is a gradient of increasing growth rate from left to right (Fig. 4.1a). In the second, the growth rate is the same throughout but growth is anisotropic (an- = not), being increasingly circumferential (transverse) toward the right-hand end and increasingly longitudinal toward the left (Fig. 4.1b). Similarly, leaf initiation can involve mainly a local increase in growth rate to form a bulge (e.g. *Silene*), or mainly changes in growth direction (e.g. *Pisum*) (see section 4.3).

In tip growth of fungal hyphae, root hairs, and moss and fern **protonemata**, the bending of the filament can be either by differential growth rates just behind the tip, i.e. bowing (Fig. 4.1c), or by bulging, i.e. essentially the establishment of a new tip to one side of the old one, analogous to branching but where the original axis ceases growth (Fig. 4.1d).

The mechanism by which shape changes often cannot be inferred easily. Only direct measurements of markers on the surface can show what the bases for change of shape are in any particular growing structure.

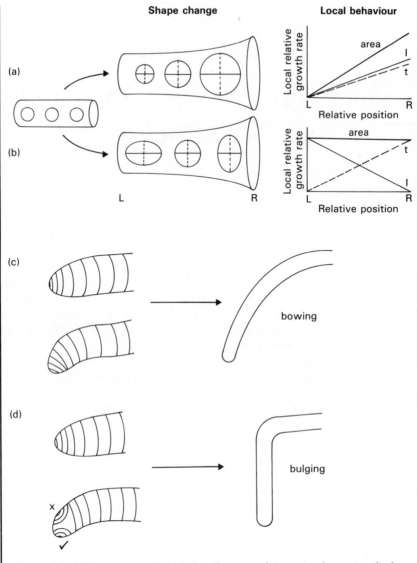

Figure 4.1 Different ways of achieving the same changes in shape. A cylinder growing to a trumpet shape could grow either (a) isotropically but with increasing area growth rate from left to right or (b) anisotropically, with the same area growth rate throughout but with a decreasing longitudinal (l) and an increasing transverse (t) component from left to right. Bending in tip growth could be either (c) by bowing, the growing point remaining at the tip but differential growth behind the tip, or (d) by bulging, where a new growing point takes over and directs growth downward. (After Green *et al.* 1970).

arrangement could perhaps be the result of mechanical forces caused by turgor pressure acting on an outer restraining layer, which is perhaps only the thick, outer epidermal cell wall. The slower growth at the apical summit and the initiation of primordia only on the flanks of the apex implies some sort of restraint on the surface, which prevents it bulging out to form primordia at the apex summit. The different structure of the root and shoot meristems suggests that the restraint in the root may diminish as a function of distance from a maximum at the tip of the meristem, whereas in the shoot apex the restraint may be similar over the whole of the surface of the apical dome. This might be expected because of the root cap being a more localized restraint. The small roots on the plantlets that form on the leaf margins in *Bryophyllum* have no, or hardly any, root cap. Displacement of marks on the surface of these roots shows that there is a higher rate of surface expansion at the extreme root tip than there is in the broadening region just behind it, unlike roots with caps. The movement of marks over the whole of the root surface also suggests that there is no quiescent centre in these roots.

The directions of growth can be observed directly on the surface of the shoot apex because it is not obscured by a cap. There is predominantly longitudinal growth, as shown by the displacement of small pieces of black marker placed on the apex surface and by the predominantly transverse orientation of cell division (normal to the direction of growth) in the epidermal cells. For a hemispherical dome, only a longitudinal polarity of surface growth is consistent with a minimum rate of growth at the summit of the apical dome, a maximum on the flanks, and a longitudinal polarity of growth in the axis below the meristem (Box 4.2).

Box 4.2   Growth gradients and meristem shape

The shape of a model hemispherical shoot apex could in theory depend on any one of three different growth distributions (Fig. 4.2). The growth could be isotropic. This would require a maximum growth rate at the summit of the hemisphere, falling to zero at the base (Fig. 4.2a). Growth could be anisotropic and predominantly transverse. This would again require a maximum growth rate at the summit falling to zero at the base. Growth of the apex could be visualized as a succession of expanding annuli originating at the apical summit (Fig. 4.2b). Growth could be anisotropic but predominantly longitudinal. This would require a minimum growth rate at the summit, a maximum on the sides of the hemisphere, and falling to zero again at the base. With increasingly longitudinal growth toward the base, so that the transverse component would be zero at the base, the growth rate could be maintained at the base, being translated into longitudinal extension (Fig. 4.2c). Only the latter is consistent with what is observed: minimum growth rate at the summit of the meristem, maximum on the flanks, and longitudinal extension at the base of the apical dome.

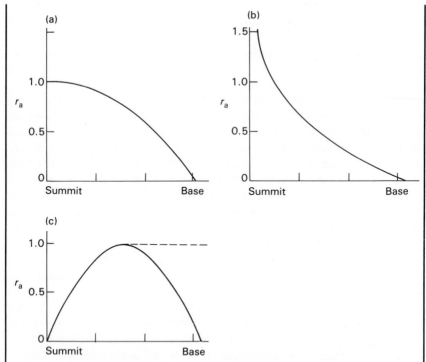

Figure 4.2 Surface growth rates and directions to maintain the shape of a hemispherical apex. (a) Isotropic growth. The area relative growth rate, r_a, falls from a maximum at the summit to zero at the base of the apex. (b) Anisotropic growth, where the transverse component is twice the longitudinal component. Growth rate falls from maximum at the summit to zero at the base. (c) Anisotropic growth, where the longitudinal component is twice the transverse component. Growth rate rises from zero at the summit to a maximum half-way down the hemisphere, then falls to zero again at the base. With an increasing longitudinal component (dashed line), growth rate can be maintained, but at the base it is entirely longitudinal. Only (c) corresponds to observation on shoot apices. (After Green 1974)

The root tips of *Bryophyllum* plantlets have a region near the tip of the meristem in which transverse growth seems to predominate (Fig. 4.2b), but just behind this the growth orientation changes to longitudinal. These roots may, therefore, show both transverse (Fig. 4.2b) and longitudinal (Fig. 4.2c) anisotropy but at different points in the growing root. More data are needed to be sure.

Tip growth, such as in fungal hyphae, root hairs, and pollen tubes, is characterized by a maximum growth rate at the tip of the growing organ, decreasing to zero at the base of the growing tip, as in Fig. 4a.

4.2 THE STRUCTURAL BASIS OF AXIALITY[2]

The simplest growing axis is that of a root or a stem, which is essentially an elongating cylinder. In cylinders filled with liquid under pressure, as turgid cells are, the force exerted on the walls would tend to burst open the ends of the cylinder, as it does in a can of lemonade. In infinitely long cylinders, the force exerted on the side walls assumes major importance. The columns or files of cells that make up the root or shoot axis can be regarded as very long cylinders, which happen to be partitioned by thin cross walls making no essential contribution to the cylinder's mechanical strength. In the root and the stem, under pressure from turgor, the stress on the side walls of the cells is contained by hoop reinforcement, the hoops being of transversely orientated cellulose **microfibrils** in the cell walls. Because the side walls are restricted in lateral expansion, growth is therefore by elongation. Since each individual cell is hoop reinforced in this way and is bonded to its neighbouring cells by the middle lamella of the cell wall, the whole organ shows net hoop reinforcement and tends to elongate. The general pattern of cellulose alignment showing hoop reinforcement in stems and roots can be seen with the aid of polarized light microscopy (Fig. 4.3) (Box 4.3). The generally transverse orientation

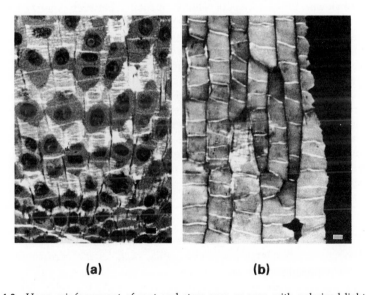

(a)   **(b)**

Figure 4.3 Hoop reinforcement of root and stem axes as seen with polarized light. (a) Longitudinal section of a *Sprekelia* root in which the cell files are sectioned obliquely so that bands of cytoplasm with nuclei alternate with cell wall. The cell walls are lighter than background, showing that they have an overall transverse orientation of cellulose microfibrils. (b) Epidermal cells of *Graptopetalum* stem are mostly lighter than background, again indicating predominantly transverse cellulose microfibrils. (Green 1984; photographs kindly supplied by Professor P. B. Green)

of the cellulose microfibrils in individual cells can also be seen by electron microscopy (Fig. 4.4). The orientation of the wall microfibrils is mirrored by the orientation of the microtubules (MTs) in the cytoplasm on the inner side of the plasma membrane, which are therefore thought to be involved in some way in the alignment of the wall microfibrils (see Ch. 5, section 5.6.1).

Box 4.3 Polarized light and the investigation of wall structure[3]

The microfibrils of the cell wall contain crystalline cellulose. This is birefringent, i.e. it has two refractive indices. The refractive index to light vibrating parallel to the long axis of the cellulose chains (and microfibrils) is greater than that to light vibrating at right angles. Light that is plane polarized, by passing it through a polarizer (like the glass in Polaroid sunglasses), will not pass through a second polarizer orientated at 90° to the first. If a piece of birefringent material is placed between such crossed polarizers, so that its axis of major or minor refraction is 45° to the planes of polarization of the polarizers, then the light passing through it becomes elliptically polarized and becomes able to pass through the second polarizer. The birefingent material between the polarizers then appears bright to the viewer. When a sheet of cellulose is rotated between polarizers at 90° to each other, then when the cellulose microfibrils are at 45° to the planes of polarization of the polarizers they will appear brightest. The cell wall material to be examined is placed on a slide and viewed down a microscope that has the first polarizer in the light path that reaches the slide and the second polarizer between the objective and the eyepiece. By the use of a Red I plate or ¼ wave plate, the axis of major refraction can be found, corresponding to the direction of the cellulose chains. To cut off sheets consisting of the outer epidermal cell walls from shoot apices requires technical skill. P. B. Green does this routinely and so has been able to provide material from which the averaged microfibril orientation in the surface walls of shoot apices, leaves, roots, and stems has been measured (Figs 4.3 & 9).

Wall structure in relation to extension has been studied in the alga *Nitella* because it has large coenocytic cells, each several centimetres long, so that it is possible to obtain pieces of cell wall large enough for study. The inner layers of the wall are hoop reinforced. As the cell extends, new layers are continually being laid down on the inside of the wall. The outer wall layers become progressively stretched and elongated as they become older and are displaced further to the outside of the wall. Eventually, the cellulose microfibrils in the oldest, outermost layers become predominantly longitudinal. There is, therefore, a gradual

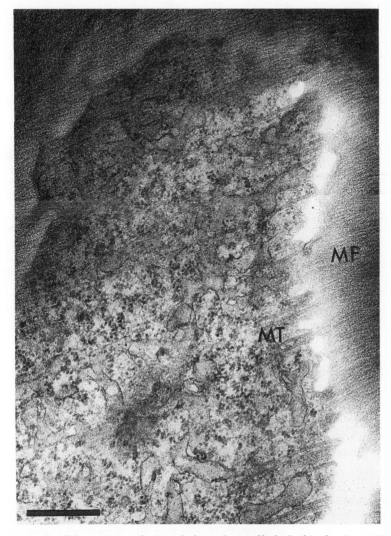

Figure 4.4 Parallel orientation of microtubules and microfibrils. In this glancing section of a *Graptopetalum* cell wall, the microfibril (MF) orientation in the wall is the same as that of the microtubules (MT) in the cytoplasm. Bar = 0.5 μm. (Hardham *et al*. 1980; photograph kindly supplied by Dr A. R. Hardham)

change in microfibril orientation from predominantly transverse on the inside of the wall to longitudinal on the outside (Fig. 4.5). This is consistent with the multi-net hypothesis of cell wall growth, which states that new wall microfibrils are predominantly transverse and that the older microfibrils become longitudinal as a result of the passive stretching of the cell wall. In the presence of colchicine, which disrupts

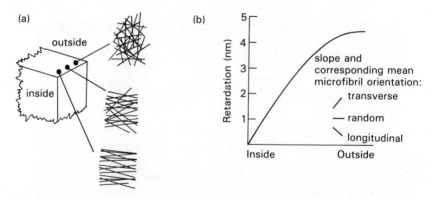

Figure 4.5 Multi-net wall growth in *Nitella*. (a) The wall microfibrils are transverse on the newly synthesized inner surface of the wall and become increasingly random further out in the older layers, which have grown for longer and have become stretched. (b) The change in microfibril orientation in (a) is shown by the slope of the curve of retardation of polarized light as measured with an interference microscope. (After Gertel & Green 1977)

microtubules, the orientation of the new cellulose microfibrils in *Nitella* is random. When the cells are left in colchicine long enough for the inner 25% of the wall to have random microfibrils, the cells tend to round up and burst, showing that it is the inner 25% of the wall that takes most of the strain and it is, therefore, the orientation of the microfibrils only in the inner wall layers that determines the axis of cell elongation. This was confirmed by the removal of the colchicine, which allowed resumption of directed growth after new wall growth, amounting to 25% of the total wall thickness.

Experiments with *Nitella* showed that the transverse (hoop) orientation of the newly-deposited wall microfibrils was maintained whatever type of strain the cell was subjected to – whether it was squashed so that it could only grow sideways or pulled so that it could only elongate. Only when the cell was constrained in a box, so that it could not grow in any direction, was the transverse pattern of microfibril deposition abandoned; deposition became random. The conclusion was that strain was necessary for ordered deposition of microfibrils but that the direction of deposition, transverse, was not determined by the direction of strain. Nor is the direction of deposition determined by the orientation of existing microfibrils. Cells in colchicine formed random microfibrils, but when the colchicine was removed transverse microfibril deposition resumed. Since the microtubules (MTs) reform in their original transverse orientation when the colchicine is removed, it is presumed that microtubule organizing centres (MTOCs) exist, which are not disturbed by colchicine and so can resume their function when microtubules are reassembled, and that MT orientation dictates microfibril orientation. Whether MTOCs are orientated in some way by strain, or indeed what they are, is unknown.

When roots are exposed to colchicine, the growing tip swells and microfibril orientation becomes random. When the colchicine is removed, transverse microfibril orientation is restored, as in *Nitella*. The elongation of cells in algae and in stems and roots of higher plants apparently depends on the hoop reinforcement of the inner layers of the cell walls, but what causes the microfibrils to become aligned like this as the shoot or root axis first forms is not known.

The transverse alignment of newly synthesized cellulose microfibrils on the inner surface of the wall is characteristic of cylindrical cells in the growing regions. When the cells mature (see Ch. 8) and stop expanding, the secondary wall that is laid down is often of a different structure, consisting of successive lamellae each with a microfibril orientation at approximately 120° to the layers on either side. This structure, like plywood, is very strong but such walls can be loosened by auxin.

4.3   FORMATION OF A NEW AXIS AND CHANGES OF GROWTH DIRECTION[4]

When a new axis forms on the side of an existing axis, and the hoop reinforcement of the original axis is also found in the new axis, then there has to be a reorientation of microfibrils from transverse to longitudinal at the sides of the site of the new axis (Fig. 4.6). How this comes about has been studied in the succulent *Graptopetalum*. A detached leaf will form a new shoot at its base from the axillary bud, which is a group of transversely elongated cells hidden in the leaf near the axil (Fig. 4.7a). The cellulose microfibrils in this meristem are orientated transversely, across the meristem and across the axis of the leaf. As the meristem begins to grow and bulge, microfibrils become orientated longitudinally across the central part of the meristem but especially at the four sites that correspond to the sides of the incipient leaves (Fig. 4.7b). These rearrangements of microfibrils (and microtubules) in the epidermis occur in conjunction with cell divisions in the meristem as the leaves are just beginning to form as slight bulges. As the leaves continue to grow, the microfibril arrangements tend to round off so that they conform to the outline of the young, more or less cylindrical leaves. The new axes so formed then possess hoop reinforcement, like the parent axis but normal to it (Fig. 4.7d).

What causes the microfibrils (and microtubules) to change orientation and so create a new strain pattern in the epidermis that allows the outgrowth of the leaf primordia to form new hoop reinforced axes? A clue comes from the cells of the underlying tissues. The first anatomical sign of the formation of a leaf primordium is often the appearance of periclinal divisions in the subepidermal cells at the potential leaf site. Periclinal

69

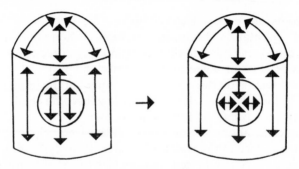

Figure 4.6 Formation of a new axis on the flanks of an old one. In the formation of a new radially symmetrical axis, there has to be a 90° shift in the polarity of growth at the sides of the site of the new axis. (After Green & Brooks 1978)

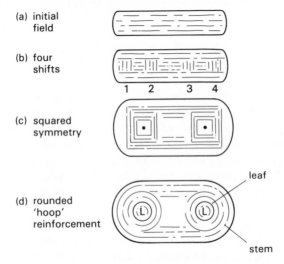

Figure 4.7 Changes in surface structure in the axillary meristem of *Graptopetalum*. (a–d) Predominant orientations of epidermal microfibrils. (a) The initial field is transverse. (b) 90° polarity shifts occur at four positions. (c) Squared symmetry. (d) Rounding off gives the two leaf sites. (Green & Poethig 1982)

divisions appeared in the *Graptopetalum* axillary meristem just as the leaf primordia began to form. The orientation of microtubules in non-epidermal cells was found to alter only in cells that had already undergone a periclinal division, but not necessarily in all of them. Division in a new plane seemed to be a prerequisite for reorientation of the microtubules. Further study of the epidermal cells led to the conclusion that the shape of the cells was also a factor in determining whether or not microtubule orientation changed (see Ch. 5, section 5.7). If the long axis of new daughter cells was parallel to the cell plate just formed between

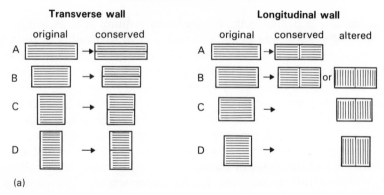

(a)

Figure 4.8 Reorientation of wall microfibrils. New microfibrils (and microtubules) are predominantly transverse to the plant axis and, with transverse divisions, they remain so. Microfibril orientation changes only after a change in division plane, to longitudinal, and only if the aspect ratio (length/breadth) of the daughter cell is about 0.7 or more. (Green 1984)

them, then the microtubule orientation altered to become also parallel to the new cell plate. Microtubule orientation did not change if the new daughter cells were broader and less elongated, with an aspect ratio (length/breadth) of 0.7 or more (Fig. 4.8). It seems, therefore, that a change in the plane of cell division and in the cell axis is necessary for the orientation of the microtubules, and hence the microfibrils, to change. The new reinforcement pattern then set up is envisaged as governing the site and shape of the new axis.

This interpretation poses two further questions:

(1) Why does the plane of cell division change?
(2) What determines the sites at which this occurs?

Plainly, for changes in cellular reinforcement fields to be effective the cells must be able to grow, and for this to happen the cell walls must be able to extend and must, therefore, be plastic. This was shown for *Nitella* cells constrained by a cylindrical glass jacket with a hole at one side. A new axis formed by growth of the cell wall out through the hole.

In meristems, the problem then is how greater plasticity is conferred on cells at certain sites and not others. This may be the primary event, and the changes in microfibril orientation may be a secondary event determining the shape of the new axis that is about to form. This is suggested by what happens in the pea (*Pisum*) shoot apex. Here, at the site of leaf initiation, the plane of cell division in the epidermis and presumably also of hoop reinforcement of the cells remains predominantly transverse and does not change as obviously as it does in *Graptopetalum*. In the pea, the predominantly longitudinal orientation of growth in the apical dome persists into the leaf primordium. It is as though the surface of the apex was being pinched and drawn out. On the other hand, in *Silene* the orientation of divisions changes, especially at the leaf site, the whole of which, and not just the sides, becomes a region of longitudinal divisions (Table 4.1). In each case the changes in division planes, and presumably of reinforcement patterns, are different but they can be related to the general shape of the emerging leaf. In *Pisum*, the young leaf is strongly dorsiventral and preserves the longitudinal polarity of growth; *Silene* is at the other extreme and each leaf at first extends laterally around the sides of the apex; *Graptopetalum* is intermediate in having leaves that at first are radially symmetrical. The changes in division planes and reinforcement pattern may, therefore, be more concerned with the regulation of the initial shape of the new axis rather than whether it forms in the first place. However, there is good evidence that the site of a new leaf may depend on the microfibrillar reinforcement pattern of the epidermis (see below). This suggests that the pattern may determine not only *where* an organ or axis may form, and its shape once formed, but also *whether* it forms.

The shape of the leaf margin may be directed by the surface structure, as shown by P. B. Green's observation in all leaves so far examined, of a band of cells running transversely across the leaf site, in which the predominant orientation of the wall microfibrils is always longitudinal to the parent axis. This band consists of those cells that form the edges and tip of the new leaf (see Fig. 4.10). Exactly what role these cells may have in determining leaf shape is not yet known.

4.4   SURFACE STRUCTURE AND PRIMORDIUM FORMATION

4.4.1   *Leaf arrangement: phyllotaxis*[5]

Leaf primordia form at discontinuities in the cellulose reinforcement pattern of the epidermis of the shoot apex. Examination of the cellular pattern of the epidermis of shoot apices, and measurement of the averaged orientation of the cellulose microfibrils in the epidermis by the use of polarized light (Figs 4.9 & 10), has shown that the microstructure of the apex epidermis is related to the sites of leaf initiation. In whorled phyllotaxis, the orientation of the apical structure changes through 90° (when leaves are in opposite pairs) at the initiation of each successive pair of leaves (Fig. 4.9). In Fibonacci (spiral) phyllotaxis, there are sectors of the apex each subtended by an existing primordium, in which the cell files seem to be parallel in the radial dimension and link the leaf base to the apex by a chevron (V) of cells in which the point of the V is towards the apex. The microfibril orientation, as shown by polarized light, in

Table 4.1   Orientation of growth in the epidermis at the site of leaf initiation just before (A) and during (B) leaf primordium formation. The main axis of growth is normal to the plane of cell division, so that in *Pisum*, where 2/3 of divisions are transverse, growth is always predominantly longitudinal. During leaf formation in *Silene*, most divisions are longitudinal, associated with transverse growth at the leaf site. (*Pisum*: unpublished data. *Silene*: Lyndon & Cunninghame 1986)

		Pisum		*Silene*	
		A	B	A	B
no. of divisions[a]	longitudinal	21	4	14	72
	transverse	38	8	15	29

[a] Data from several apices in each case combined.

these Vs of cells is tangential, which is consistent with the radial expansion of the apical surface (Fig. 4.10). Green points out that the sites of new leaves lie on the discontinuities where two V-shaped sectors of cells abut. These are regions of rapid (but not always discontinuous) changes in reinforcement curvature. He believes that the V-shaped sectors and the tangential orientation of the microfibrils are the result of the tangential growth of the primordia and, hence, their stretching action on the apex once they are initiated. Thus, the sites of future leaves are prescribed by the arrangement of the existing leaves, especially those that are 3 and 5 **plastochrons** old at the time of initiation of the new primordium. This interesting theory, based on the new evidence of apical microstructure, is able to predict the sites of primordia in a steady-state system that is already functioning. It has not yet been addressed to the problems of how the transition from one phyllotactic system to another might occur, as happens in normal growth, for instance when phyllotaxis changes from opposite to spiral or when spiral phyllotaxis is established in the seedling, but starts with opposite cotyledons. This would seem to involve a change from primordia being initiated at potential sites that are rounded off in whorled phyllotaxis to sites arising at discontinuities in spiral phyllotaxis (Figs 4.9 & 10).

Primordia do not always form at sites that are precisely predictable, especially in spiral systems. In *Sinapis* and *Chrysanthemum*, for example, the divergence angles of successive leaves may vary considerably and the Fibonacci angle of 137.5° is only a mean value that is approached. Perhaps this argues for the greater importance of the self-regulatory inhibitory fields in spiral phyllotaxis but for a greater input from the surface structure in the precise positioning of the whorled systems. What seems most likely is that morphogen fields and surface structure interact in causing primordium initiation and that neither alone is solely respon sible. However, it is easier to envisage changes in morphogen

(a)

(b)

these sectors change orientation

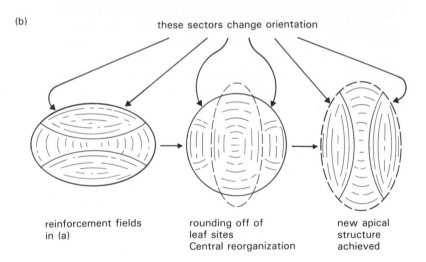

reinforcement fields
in (a)

rounding off of
leaf sites
Central reorganization

new apical
structure
achieved

Figure 4.9 Decussate (whorled) leaf formation in *Vinca*. (a) The isolated outer epidermal surface of the shoot apex in polarized light shows lighter regions where reinforcement is north to south and darker regions with reinforcement east to west. Note the corridor of reinforcement at 90° across the centre. The next pair of leaves will form at east and west. (b) The changes in reinforcement field are primarily in the four 'corner' sectors and result in paired regions where rounding off occurs to give the leaf sites. (Green 1985; photograph kindly supplied by Professor P. B. Green)

(a)

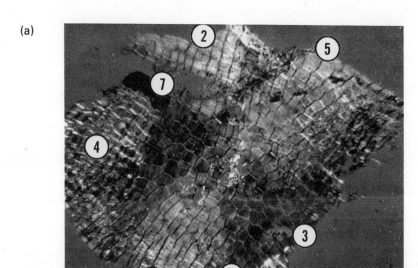

(b)

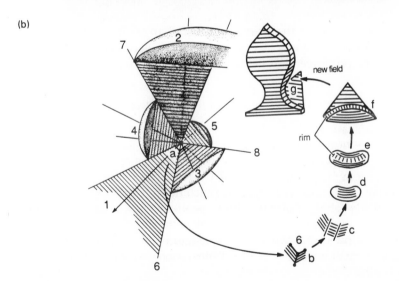

Figure 4.10 Spiral (Fibonacci) leaf formation in *Ribes*. (a) The isolated outer epidermal surface of the shoot apex viewed by polarized light has V-shaped sectors of different reinforcement orientations, which converge on the centre. (b) The next leaf (6) arises at the discontinuity in the reinforcement field where the angular contrast is greatest. The shift in polarity across the leaf site is in the cells that form the leaf margin. (Green 1985; photograph and diagram kindly supplied by Professor P. B. Green)

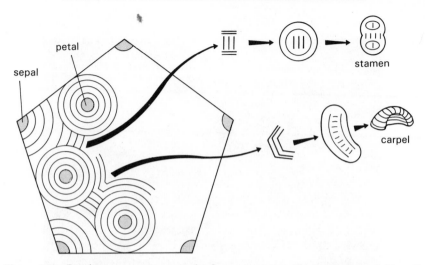

Figure 4.11 Reinforcement patterns in the flower meristem. The petals form at discontinuities in the reinforcement field associated with the sepals. Stamens form at sites where rounding off can occur, where sepal and petal fields interact and give organs with a row of cells over their summits showing transverse reinforcement. Carpels form at internal discontinuities in the files produced by the other organs. (After Green 1988)

concentrations as being the prime cause of changes in epidermal wall plasticity, which then allow changes in division planes and consequent changes in the reinforcement fields.

4.4.2 Arrangement of organ primordia in the flower[6]

When flowers are formed, the ordered structure of the shoot apex changes and whorls of primordia are formed. The formation of a new reinforcement pattern of microfibrils in the epidermal cell walls of the flower meristem has been described by P. B. Green for *Crassula* and *Echeveria* (Fig. 4.11). The five sepals are each hoop reinforced and, where adjacent fields abut, polarity shifts produce new potential hoop reinforcement sites, which form the petals, alternating with the sepals. The stamens form at sites where the sepal and petal reinforcement fields combine to form new hoop-reinforced sites, each with a radial file of cells over its centre.

Floral organ primordia very often arise in helical sequence, just as leaves do in Fibonacci phyllotaxis. The sequence in which the primordia arise, however unusual it may sometimes be, must also be under the control of the apex. An attractive possibility is that, in the flower, the surface structural features are concerned primarily with the positions of the primordia and that an inhibitor (or promoter) field is more concerned with the local realization of this potential and the timing of it. It may be that the sequence is first determined by an inhibitor type of mechanism, and this results in surface strains that orientate the microfibrils, which in turn then influence subsequent primordial positioning. In leaf positioning in spiral phyllotaxis where sequence and position coincide, the two postulated systems would work to reinforce each other. In the flower, they may act non-simultaneously.

In the development of some bilaterally symmetrical, zygomorphic flowers, as in some of the Leguminosae, the primordia are not formed in the expected sequence that would be deduced from the mature flower. Instead, the primordia form first toward the front of the floral apex and then primordium formation spreads in a wave to the back. Different types of floral organs are formed simultaneously. This unusual form of development shows that the type of organ formed and its position are not determined by the sequence of development; rather, it appears as though a preformed pattern becomes realized. In other flowers, the exact positioning of one whorl of organs, not only with respect to the other organs of that whorl but also in relation to the other whorls, is difficult to reconcile with an inhibitor type of mechanism as postulated in the field theories of phyllotaxis. The different floral whorls seem to have a promotory rather than an inhibitory effect on each other's positioning.

The problem still remains of understanding how the patterning process changes. Again, this may be less difficult to understand if there are two interacting systems. If (hypothetical) inhibitors produced by the newly initiated primordia were to reduce in concentration or rate of production on flowering, especially if one of these substances resembled auxin and rendered the surface locally more plastic, so controlling the area of a primordium at initiation, this could perhaps result in the reduction in primordial size in the flower (see Ch. 3, Table 3.3). The associated change in primordium arrangement would also presumably then alter the strain pattern in the epidermis. This could then allow the transition from the reinforcement pattern during leaf initiation to that in the flower.

4.5 SUMMARY

(1) Changes in shape can be brought about by local changes in growth rates or changes in directions of growth, or a combination of both. Only direct measurements can show what is happening. The increasing growth rate from the summit of a hemispherical shoot apical dome downward is necessarily associated with a predominantly longitudinal anisotropy of growth.

(2) Plant axes characteristically show hoop reinforcement of their cell walls by the component cellulose microfibrils. This restricts lateral

expansion and favours longitudinal extension. The cellulose micro-fibril orientation is paralleled by the orientation of the cytoplasmic microtubules, which are therefore believed to control in some way the microfibril arrangement. In the alga *Nitella*, the cell walls undergo multi-net expansion. The initially transverse microfibril orientation depends on the existence of strain but this does not seem to be the orientating factor. Secondary walls of higher plant cells may be of crossed polylamellate structure.

(3) The formation of a new hoop-reinforced axis on the side of a parent axis requires a 90° shift in microfibril orientation at the sides of the new axis site. The microfibril pattern then rounds off to give a circular reinforcement pattern, which becomes the transversely aligned hoop reinforcement of the new axis. Polarized light has been used to show that, in higher plants, the necessary changes in cellulose microfibril orientation occur after changes in cell division planes or changes in cell shape as a result of differential rates of extension and division.

(4) Changes in division planes and epidermal microfibril orientation during leaf initiation are consistent with the reinforcement patterns being concerned with the shape of the newly initiated leaf prim-ordium. The rim of a leaf has a distinct pattern of microfibril orientation normal to the plane of the leaf blade, which otherwise shows essentially hoop reinforcement. Whether these marginal cells have a determining role in leaf shape is not known.

(5) In spiral phyllotaxis, new leaves arise at regions of sharp curvature, sometimes discontinuities, in the microfibril reinforcement field of the shoot apex. Each leaf subtends a radial field of tangentially reinforced cells. In whorled phyllotaxis, the leaves form at positions in the reinforcement field that lie between regions of different alignment and can produce rounded-off regions, as in *Graptopet-alum*. Also, in flowers, floral organs form at sites where reinforce-ment fields of existing primordia interact. The epidermal cell reinforcement fields seem to be the basis for primordial positioning. Interacting with them may be morphogen fields, concerned also with primordial positioning and also with the sequence in which primordia arise, which, even in the floral whorls, is often in spiral sequence.

FURTHER READING

Green, P. B. 1986. Plasticity in shoot development: a biophysical view. *Symposia of the Society for Experimental Biology* **40**, 211–32.

Green, P. B., R. O. Erickson & P. A. Richmond 1970. On the physical basis of cell morphogenesis. *Annals of the New York Academy of Science* **175**, 712–31.

Robards, A. W. (ed.) 1985. *Botanical microscopy 1985*. Oxford: Oxford University Press. (Chapters on cell wall structure, synthesis, and the cytoskeleton)

NOTES

1 Green (1974). (*Bryophyllum* rootlets; shape and growth gradients)
2 Gertel & Green (1977). (*Nitella*)
3 Preston (1974). (Polarized light microscopy)
4 Green & Brooks (1978), Green & Poethig (1982), Lyndon & Cunninghame (1986). (Formation of a new leaf axis)
5 Green (1985). (Wall structure and leaf positioning)
6 Green (1988), Tucker (1984). (Flower formation)

CHAPTER FIVE

Control of the plane of cell division

5.1 CELL DIVISION AND ITS RELATION TO THE GROWTH AXIS[1]

The growth and elongation of organs depends on the growth and elongation of the component cells. Growth is accompanied by cell division, except in the final stages of cell expansion during the later stages of growth of leaves, flowers, and fruits and in the proximal parts of the growing zones in stems and roots. In meristems, cell division roughly keeps pace with cell growth so that cells remain approximately the same size. Cell shape also usually remains relatively unchanged because the plane of cell division is usually normal to the predominant direction of growth.

In meristems, most cell divisions are proliferative, i.e. they simply increase cell number, in contrast to the formative divisions where the daughter cells give rise to cell lineages that differentiate from each other. The proliferative divisions in the root are transverse to the cell files. The formative divisions are longitudinal to the cell files and so increase file number. In the root, these divisions are mostly confined to the tip where it is expanding laterally and so these longitudinal divisions are also normal to the predominant direction of growth in that part of the root where they occur.

In the shoot, as in the root, in the elongating region the divisions are predominantly transverse and normal to the direction of elongation. The formative divisions giving rise to the procambium and other elongated (usually vascular) cells seem to occur not by longitudinal divisions in the elongating region itself but in the shoot apex. Here, longitudinal division is normal to the predominantly transverse growth axis just below the apical dome where the stem begins to widen. In the shoot apex, the formative divisions that give rise to nodes and internodes will be transverse to produce cell layers giving rise to the successive regions of the stem axis. In the root, there is no such subdivision of the axis modules as in the shoot. The elongated shape of procambial and vascular cells is because of their elongation without corresponding transverse divisions. However, elongate cells such as those of the cambium may divide essentially longitudinally if the main axis of growth is across the stem, e.g. when the stem is increasing in girth during **secondary thickening**.

In the cambium, cell shape is determined originally in the procambial stage or when the cambial cells arise from interfascicular parenchyma cells. These latter are usually much shorter than the cambial cells formed from them, implying that the cambial cells have lengthened after formation. They can do this by intrusive growth, in which the tips of the cell grow up or down between the other cells. The pseudotransverse (anticlinal) divisions that later extend the circumference of the cambial cylinder are essentially normal to one of the two axes of growth – the tangential direction. The other (periclinal) divisions of the cambium are normal to the radial growth axis. The relative rate of tangential expansion, in circumference, is necessarily always exactly equal to the relative rate of radial expansion and so we might expect equal frequencies of periclinal and anticlinal divisions in the cambium. Because of the difficulty of measuring the rate of periclinal division in the cambial initials themselves, it is not known what the ratio of periclinal to anticlinal divisions is. There is no reason yet to believe that the cambium does not obey the apparent rule that the plane of cell division is usually normal to the principal direction of growth. There is, however, evidence that the plane of division, especially in the cambium, is strongly influenced by pressure (see section 5.4.2).

A study of cucurbit fruits showed that, irrespective of the actual shape of the fruit, some of which were long, some broad, and some isodiametric, the predominant plane of cell division was normal to the main axis of growth in every case. In the isodiametric fruits with growth in all directions, the cell divisions were also in all planes.

In developing leaves, the planes of cell division have also been shown to be normal to the predominant growth axes. In *Myriophyllum* leaves, which form many leaflets, the divisions are normal to the long axis of growth of the whole leaf in the sinus regions where outwardly directed growth is absent or minimal but normal to the lobe axis in or beneath the lobes themselves (Fig. 5.1). In *Tropaeolum* leaves, growth rates were measured by marking leaf surfaces and again the predominant plane of cell division was found to be normal to the direction of fastest growth.

The general rule is that the plane of cell division is normal to the growth axis, or the frequency of divisions in different planes is proportional to the rates of growth normal to these planes. Possible exceptions, such as the occurrence of periclinal division without obvious outward growth in the *Silene* shoot apex (although the frequency of anticlinal

(a)   (b)   (c)

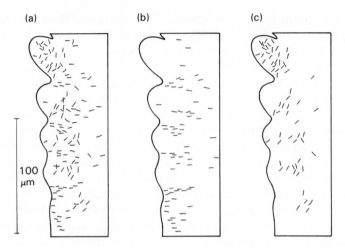

Figure 5.1 Division planes in developing *Myriophyllum* leaves. (a) Cell division planes in part of a young leaf in which lobes are forming tend to be normal to the locally predominant axis of growth. (b) In the sinus regions, the divisions are mainly transverse to the leaf's long axis. (c) Where the lobes form, divisions tend to be normal to the axis of the lobes. (Jeune 1975)

divisions was not measured) and the formation of cambium-like layers under wounded surfaces, may be accounted for by assuming that in these cases the overriding factor is pressure and that the plane of cell division is normal to the principal direction of stress (see section 5.4.2).

5.2 ARE CHANGES IN THE SHAPE OF PLANT PARTS CORRELATED WITH DIFFERENTIAL RATES OR PLANES OF CELL DIVISION?[2]

The change in shape of any organ could theoretically result from local changes in growth rate (an increase causing a bulge or a decrease forming a hollow or indentation), or from changes in the directions of growth, or from a combination of both. The direction of growth can be inferred from cell shapes and planes of cell division. If the cell shape in a growing organ remains approximately constant, this means that the direction of growth is basically normal to the plane of division, otherwise cell shape would change. If cell shape remains unchanged but there is no preferred plane of division, this implies that growth is isotropic, i.e. equal in all directions, and any change in organ shape must therefore be by localized changes in the rate of growth. In higher plants, the shape of the root, stem, leaves, and fruit in the examples studied seems to be correlated with the main plane of cell division being roughly normal to the main growth direction. However, the mature shape of leaves, for example, is

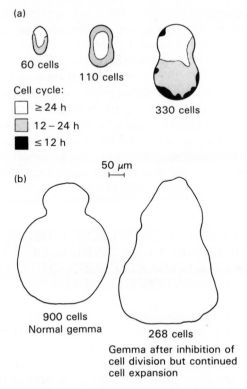

(a)

60 cells

110 cells

Cell cycle:

☐ ≥ 24 h

▨ 12 – 24 h

■ ≤ 12 h

330 cells

50 μm

(b)

900 cells
Normal gemma

268 cells
Gemma after inhibition of
cell division but continued
cell expansion

Figure 5.2 Development of lobes in gemmae of *Riella* (liverwort). The gemmae are undifferentiated asexual reproductive structures. (a) The lobes are apparently formed by a locally greater rate of cell division (shorter cell cycle) in the lobe margins than elsewhere. (b) When cell division is inhibited but cell expansion continues, the shape does not change as it does in the untreated gemma, consistent with cell growth being isotropic and gemma shape resulting from differential rates of cell division. (After Stange 1983)

at least partly determined by the rates of growth. Cell division persists longest at the base of the leaf, which is usually the widest part. The tapered dicotyledonous leaf matures first at its tip and maturation and cessation of cell division passes in a slow wave down the leaf (see Ch. 8, section 8.2.1). The grass leaf is not fundamentally different. Maturation also progresses from the tip but growth at the base of the grass leaf continues because meristematic activity here persists much longer than it does in the dicotyledonous leaf. The linear shape of the grass leaf is because the main direction of growth is axial; the plane of cell division is mostly transverse to the axis.

In the embryo, change in shape may be brought about partly or primarily by different rates of growth e.g. in the formation of the cotyledons, where the rate of division is higher than in the adjacent axis. The growth of the lobes in the fern **prothallus** also seems to be mainly because of the slowing of growth in the notch region and its persistence elsewhere. Similarly, the formation of lobes in the gemmae of the liverwort *Riella* is because of the higher rate of growth and cell division in the lobe areas than elsewhere in the **thallus** (Fig. 5.2). When mitosis was inhibited, the shape of the thallus was altered because cell enlargement continued, but isotropically. In higher plants, the scarcity of examples of shape being determined by rates of growth suggests that, usually, shape is determined by the directions of growth. Where shape is determined by gradients in the rate of growth, and there is no preferred direction of growth (growth therefore being isotropic), the plane of cell division may either be random or, if there is a predominant growth direction for individual cells, could still be normal to it.

5.3   IS THE DIRECTION OF GROWTH DETERMINED BY THE PLANE OF CELL DIVISION, OR IS THE PLANE OF CELL DIVISION CONSEQUENTIAL ON THE DIRECTION OF GROWTH?[3]

When the direction of growth and the axis of the mitotic spindle coincide, i.e. the plane of cell division is normal to the growth axis, it looks as if the direction of growth depends on the plane of division, but it is the opposite that is probably true. This is shown by experiments that stop cell division without stopping growth. In wheat seedlings irradiated with γ-rays, cell division was arrested but growth continued for 10–12 days and the plantlets grew to a size comparable to that of 4-day-old unirradiated seedlings. Cell enlargement in the γ-plantlets was considerable and cells grew to sizes much larger than normal because all growth was by cell enlargement and cell number did not increase at all. Cell differentiation continued in the absence of division. The form of the organs that developed, and of the plantlet as a whole, was similar to that of unirradiated seedlings but the γ-plantlets formed no more organs and no new cell types, e.g. root hairs, for which formative (unequal) cell division were required. Treatment with gibberellic acid increased leaf extension to the same degree in γ-plantlets as in unirradiated plants. Treatment with colchicine (which inhibited cell division in unirradiated plants) had the effect of causing the same degree of widening of the leaves in both γ-plantlets and unirradiated plants. The polarity of leaf expansion and the shape of the leaves was, therefore, unaffected by whether or not the cells were dividing and did not depend on the number of cells in the plants (Table 5.1). The growth of the γ-plantlets was accommodated by cell enlargement, which produced cells of abnormal shape. The effect of cell division in the unirradiated normal seedlings was, therefore, to affect cell shape but not organ shape.

Table 5.1 Polarity of growth in δ-plantlets. Inhibition of cell division by δ-irradiation resulted in smaller leaves but the effects of 2×10^{-4} M gibberellic acid (GA_3) and 0.03% colchicine on shape, as shown by the polarization index, are the same in unirradiated and δ-irradiated plants. Polarization index is the allometric constant, b, in the allometric equation: $y = ax^b$, where y = length and x = width. Initial leaf length in the embryo was 0.66 mm and width was 1.23 mm. (Haber & Foard 1963)

Plants	Treatment	Leaf length (mm ±SE)	Leaf width (mm ±SE)	Polarization index
6-day-old unirradiated	none (water)	80.00 ± 1.54	3.82 ± 0.06	4.2
	GA_3	125.44 ± 3.10	3.19 ± 0.08	5.5
	colchicine	3.32 ± 0.26	5.58 ± 0.28	1.1
9-day-old δ-plantlets	none (water)	12.71 ± 0.61	2.54 ± 0.80	4.1
	GA_3	21.48 ± 0.80	2.28 ± 0.06	5.6
	colchicine	2.78 ± 0.28	4.08 ± 0.26	1.2

Although organ formation ceased in the γ-plantlets, they did in fact begin to form a new leaf primordium at the site on the shoot meristem where the next leaf would be expected. This leaf primordium consisted only of an expansion of the epidermal cells, and it developed no further. Therefore, cell division seemed unnecessary for the initiation of this primordium but was apparently essential for its further development. It could be argued that this leaf had already been initiated before irradiation and that it was simply showing limited enlargement.

The application of colchicine to roots can stop cell division without stopping growth. Lateral root primordia can continue to be initiated but there is no cell division so they do not develop (Fig. 5.3). However, if the colchicine is removed, division is resumed and these **primordiomorphs** develop into lateral roots. The primordiomorphs show that the initial stages in organ formation do not depend on cell division but on a change in the direction or axis of growth. Similar observations have been made in meristems treated with hydroxyurea, which inhibits cell division but does not immediately inhibit growth. The meristems retain their shape but the cells become elongated in the direction of growth.

All these experiments show that the plane of cell division does not determine either the direction of growth or the shape of organs. It only affects the shape of cells according to the plane in which they are divided. The problems are, therefore, to understand what controls the plane of division, to produce cells of relatively uniform shape, and how this control is modified in unequal divisions, which precede cell differentiation.

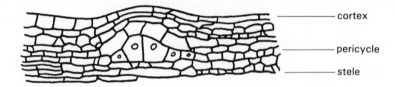

cortex

pericycle

stele

Figure 5.3 Root primordiomorph formation. When cell division in wheat seedlings was prevented by colchicine, lateral root primordia continued to be initiated by radial expansion of the pericycle cells. Only the pericycle and adjacent tissues are shown here, in longitudinal section. (After Foard *et al.* 1965)

5.4 FACTORS ORIENTATING THE PLANE OF CELL DIVISION

Observation of cell division and the resulting cell shapes and cell wall patterns has led to the formulation of rules for cell division. These rules state that the cell plate:

(1) is of minimal area for halving the volume (Errera's rule);
(2) forms normal to the main growth axis of the cell (Hofmeister's rule), which usually means that it forms at right angles to the long axis of the cell (Hofmeister's rule as it is often stated);
(3) forms at right angles to existing walls (Sachs' rule).

Proliferative divisions often seem to obey these rules; however, unequal, formative divisions often break all of them. Does cell division during general growth conform to these rules and, if so, how?

5.4.1 Cell shape and aspect ratio[4]

Since the plane of cell division is usually normal to the axis of growth, this means that cells tend to divide so that the new cell wall forms across the long axis of the cell. How is the plane of division controlled? The planes of cell division within the root tip are essentially either transverse or longitudinal, with longitudinal division being more or less restricted to that region of the meristem where the root tip broadens. Whether or not a cell in the maize root meristem divides transversely or longitudinally depends on its shape at mitosis. The aspect ratio (ratio of length/width) was 2.55 for cells dividing transversely and 1.27 for those dividing longitudinally. This significant difference ($P<0.001$) in cell shape at the onset of division implies that it is the shape of the cells that determines the plane of division. Since the cells become broader where the root tip becomes broader, this suggests that transverse growth in this region of the root is faster than longitudinal growth and that the plane of cell division is normal to the principal axis of growth. It seems to be the change in shape of the root, and the differential growth directions that bring this about, that determines the pattern of division within it. If it were the plane of division that determined cell and root shape, we might have expected the aspect ratio of all cells at mitosis to be similar.

In the shoot apex, the control of division planes is most obvious in the tunica in which, by definition, all divisions are anticlinal. The planes of division do not seem to be related to the shape of the organ as a whole, as they seem to be in the root. In the shoot, the plane of division is clearly not related to the aspect ratio because, in the tunica, cells of all widths divide anticlinally, which is more consistent with the direction of growth than the shape of the cell determining the plane of division. The shape of the cells is the result of the rate of division relative to the rate of cell growth. Surface expansion is accompanied by divisions at the shoot apex surface only in a plane normal to the growth direction, i.e. normal to the surface, whatever the shape of the cells.

Cell shape itself may not therefore be the dominant factor in determining the plane of division, but it may sometimes seem to be so when the principal growth axis coincides with the long axis of the cells. Clearly, there are other factors at play.

5.4.2 *Mechanical stress*[5]

The effect of mechanical stress in orientating the plane of division is particularly clear from the cambium. Transverse sections of wood show the cambial cells dividing tangentially, and not radially as they would if division were always across the least width of the cells. But this tangential plane of division is, of course, normal to the direction of the radial growth during secondary thickening. The role of pressure, exerted by the constraining bark, has been shown by lifting up a flap of bark. The exposed cambium then proliferates and forms a callus. If this flap of bark is replaced and fastened down, the cambium instead develops normally with ordered periclinal divisions. Cell-to-cell contact between bark and cambium is not required because the cambium resumes functioning equally well if there is a sheet of polythene between it and the bark. A cambium-like zone of cells also forms in tobacco callus in culture when this is subject to continuous mechanical pressure from a clamp (Fig. 5.4). In this callus, and in the cambium, the plane of division depends on the direction of principal stress, in these cases pressure, and is normal to it. The characteristic orientations of cell walls at the site of initiation of an axillary bud, the so-called shell zone, also look as though they could be the result of divisions normal to the directions of stress imposed by turgor as suggested by models (Fig. 5.5).

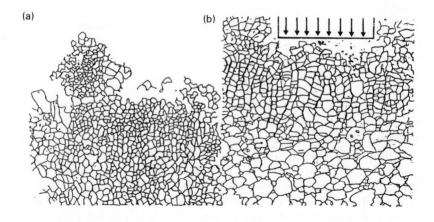

Figure 5.4 Effect of directional pressure on cell division in tobacco callus. (a) Unstressed callus shows random cell proliferation. (b) Pressure imposed locally by a clamp causes ranks of tangential cambium-like divisions orientated normal to the direction of pressure. (After Lintilhac 1984)

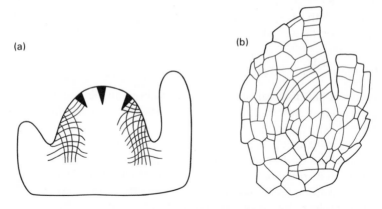

Figure 5.5 Stress patterns in shoot apices. Glycerin-jelly model sections of shoot apices show stress patterns (a) that resemble the cell division pattern of axillary bud sites, as in *Silene* (b). The cell pattern at the axillary sites could result from cells dividing in the plane of least shear. (After Lintilhac & Vesecky 1980)

There are no clear examples of the plane of cell division in plants being orientated by tension, although anticlinal division in cortical cells of stems undergoing secondary thickening and longitudinal divisions at the sides of root caps where the cap is being pulled up the side of the root may be such examples.

5.4.3   Growth substances

Only gibberellins seem to be able to affect the plane of cell division, but even this is apparently an indirect effect. Gibberellic acid (GA_3) was first noted as acting in the subapical meristem of the shoot to promote cell division and stem elongation. The bulk of the newly induced cell divisions are transverse, so that the ratio of transverse to longitudinal divisions increases and GA_3 seems to have affected the plane of division. However, the primary action of GA_3 seems to be to promote stem elongation, with increased cell division being a consequence of the increased growth. Since the plane of division will tend to be normal to the main axis of growth, the inevitable effect of GA_3 would be to result in an increased frequency of transverse divisions, but only as an indirect result of its effect on elongation. This would be consistent with the effect of GA_3 being on the polarity of growth irrespective of whether or not there was cell division, as shown by the γ-plantlets (see section 5.3).

5.4.4   Cell division in the plane of least shear – a unifying hypothesis

It is clear that Rules (1) and (2) (see section 5.4) are not followed by many cells, e.g. in cambium (Fig. 5.4) or in narrow tunica cells in the shoot apex. Because of the general orientation of the new cell wall normal to the growth axis or the direction of pressure, Lintilhac proposed that the new cell plate forms in the plane of least shear in the cell. We can think of the cell plate forming in the place in which it is least subject to distortion (Fig. 5.6). This hypothesis also means that where there is growth in more than one direction and it is neither isotropic nor entirely anisotropic, then the resultant growth vector would dictate the plane of least shear and hence the plane of cell division. This idea is equally valid when the cells are under unidirectional stress, e.g. pressure, as in the cambium, or under possible tension, as in parts of the axillary shell zone in the shoot or at the sides of the root cap. The cell wall would form normal to the direction of pressure or tension. The change in plane of cell division in the fern sporeling can also be accounted for using this hypothesis (see Ch. 6, section 6.3.2). However, some divisions of the epidermis in *Graptopetalum* are not readily explainable by this hypothesis, but this may be because the plane of least shear may not always be easy to deduce.

5.5   UNEQUAL CELL DIVISIONS[6]

All three 'rules' of cell division (see section 5.4) are most frequently broken by unequal (formative) divisions, which give daughter cells that develop differently. They often involve the cutting off of a small, densely

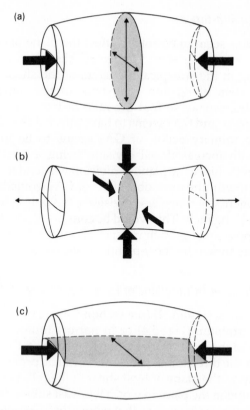

Figure 5.6 The plane of least shear. (a) The plane in which the cell plate (shaded) is least likely to be sheared because of cell compression through growth is normal to the direction of stress and the principal growth axis. (b) The plane of least shear is also transverse for a cell under axial tension. (c) A cell plate in the plane of stress would be liable to maximum shear. (Lintilhac 1974)

cytoplasmic cell from a larger, more vacuolate one, as in the formation of root hair cells or **stomata**. They also tend to break the rule that the new wall is orthogonal to those that it abuts. In the *Azolla* root, for example (see Ch. 2, Fig. 2.2), the cell walls in the formative divisions may join the existing walls at angles of 40° or less, and may sometimes be curved and not of minimal area. Unequal divisions are always found as precursors of cell differentiation; it seems that an asymmetrical distribution of cell contents between the daughter cells is necessary for differentiation. Often, before division, the nucleus moves to one end of the cell or there is already an asymmetric distribution of cytoplasm so that division results in an unequal distribution of cytoplasm to the daughter cells. This is characteristic of the divisions that give rise, for example, to sieve tube and companion cell, root hair and epidermal cell, and stomatal mother cell and epidermal cell. How is the plane (and position) of division controlled in unequal divisions?

The best-documented examples of unequal divisions are those involved in the formation of stomata. A regular sequence of divisions forms the stomatal complex (Fig. 5.7). Unequal divisions first form the guard mother cells (GMC) alternating with non-stomatal cells. The nuclei in the adjacent cells then migrate next to the GMC. They then divide unequally to give small cells next to the GMC. The lateral pair of these cells divide again and the GMC itself also divides in the same plane to form the guard cells. Occasionally abnormal divisions and the formation of extra cells also show that the positioning of the nucleus before division and the position of the new cell wall are influenced by the position of the nucleus in the adjacent GMC. Some influence seems to be transmitted across the cell walls to orientate the mitotic spindle in the adjacent cells and to determine the position of the cell plates. In *Allium*, the plane of cell division is not necessarily transverse to the spindle axis because the cell plate rotates in the cell, with the spindle and anaphase/telophase chromosomes attached, up to 90° from its original position (Fig. 5.8) so that the final cell wall is longitudinal. These divisions and the reorientation of the spindle can be disturbed by the inhibitors cytochalasin and colchicine, which disrupt microfilaments and microtubules, respectively. Presumably, the cytoskeleton is, therefore, responsible for positioning the nucleus and for determining the plane of division. However, it is far from clear what these orientating factors are in unequal divisions.

Why the divisions leading to stomatal formation are unequal in the first place, and why there is a regular sequence of changes in the plane of division are not known. The changes in division orientation do not seem to be related to changes in the axis of growth of the leaf, which, in monocotyledons remains predominantly longitudinal. As shown by the structure of the epidermis, there does not seem to be any transverse growth at stomatal sites in excess of that occurring anywhere else in the leaf, even during the stages of longitudinal divisions. Even if detailed observations of the rate of division in relation to the rates of elongation and expansion in the stomatal mother cells showed that there were subtle changes, this would still require there to be some relationship between cell shape or direction of growth and the plane of division. This is certainly not apparent from the sequences (as in Fig. 5.7) observed so far. Furthermore, the oblique orientation of the spindle in *Allium* takes place without regard to the axis of the cells or to their size. It is difficult to conclude that in unequal divisions the plane of cell division is in the plane of least shear, especially in those cases where a small cell is cut off from one corner of the mother cell. There may be an involvement of cations in positioning the nucleus, as in the germinating fern spore (see Ch. 6, section 6.4.1.3), but so far no

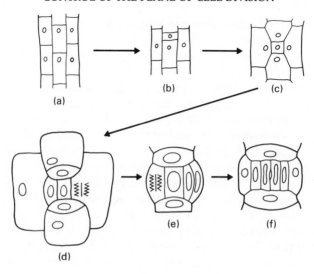

Figure 5.7 Stomatal formation in *Commelina*. (a,b) Epidermal cells divide unequally to give the small guard mother cells (GMC). (c,d) The nuclei of the adjacent cells migrate next to the GMC and divide unequally. (e) The cells lateral to the GMC divide again. (f) Finally, the GMC divides to give the guard cells. (After Pickett-Heaps 1969)

measurements of ion concentrations or of ion fluxes during the development of stomata seem to have been made.

5.6   THE ROLE OF THE CYTOSKELETON

As well as forming the mitotic spindle, microtubules have been implicated in the orientation of the wall microfibrils and, therefore, in the reinforcement patterns that correlate with the polarity of cell growth (see Ch. 4, section 4.2). Microfilaments are involved in the positioning of the nucleus before division and, in some cells, the positioning of the cell plate afterwards. The plane of cell division and the direction of growth of the cells, and hence the shape of the organs, may, therefore, ultimately be controlled by the cytoskeleton.

The plant cytoskeleton consists of the same components as the animal cytoskeleton. In plants, three main elements have been identified so far: microtubules, microfilaments, and intermediate filaments.

5.6.1   Microtubules[7]

Microtubules in plant cells form four sets of structures sequentially throughout the cell cycle:

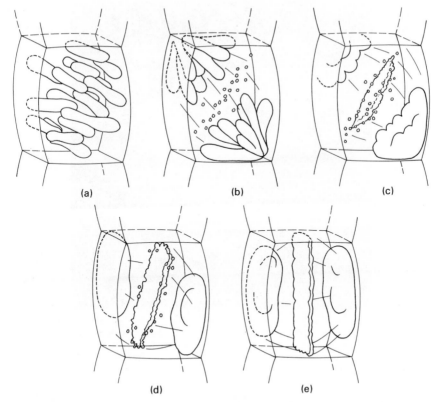

(a)   (b)   (c)

(d)   (e)

Figure 5.8 Rotation of the cell plate in *Allium*. (a) The metaphase plate lies at an angle to the cell axis. After anaphase (b), as the cell plate is forming (c), together with the nuclei, it alters position to become longitudinal in the cell (d,e). The cell plate is, therefore, moved and positioned independently of the movement and positioning of the nucleus before division. (Palevitz & Hepler 1974)

(1) Microtubules (MTs) in interphase are typically more or less transversely aligned and looped round the cell normal to the axis of cell elongation (Fig. 5.9). Before cell division, these MTs become bunched together, usually near the middle of the cell to form the . . .

(2) Preprophase band (PPB). This is a band of MTs around the cell in the position where the cell plate, formed as a result of the ensuing division, will join the existing walls. The PPB always accurately predicts the position and plane of division, even in unequal divisions. Even in *Allium*, in which the cell plate rotates (Fig. 5.8), the PPB marks the position at which the cell plate finally fuses with the side walls. In *Azolla* cells where the division is unequal and the cell plate is curved, the PPB is also curved along the exact position that the new cell wall will occupy. As mitosis is approached, the PPB disappears and is replaced by the . . .

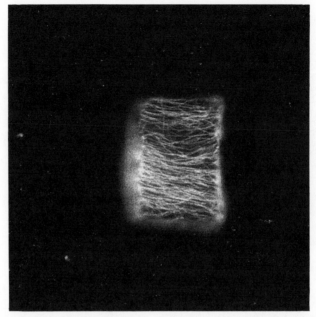

Figure 5.9 Structure of the microtubule cytoskeleton in an onion root tip cell. The microtubules are made visible because they bind antitubulin antibodies that have been tagged with a fluorescent molecule. The microtubules are roughly transverse to the cell's long axis, giving a typical 'shredded wheat' appearance. (After Lloyd *et al.* 1985; photograph kindly supplied by Dr. C. W. Lloyd)

(3) Mitotic spindle. The MTs of the spindle are orientated longitudinally in the cell. By metaphase, some spindle elements are attached to the chromosomes at the kinetochore, which is a region of the chromosome at which MT disassembly occurs. Since the other ends of the MTs are apparently anchored at the poles of the spindle, the disassembly of the MTs at the kinetochore, while still remaining attached to the chromosome, causes the chromosome to move to the spindle pole. The rate at which MT disassembly at the kinetochore takes place *in vitro* corresponds to the rate at which the chromosomes move to the poles *in vivo*. At telophase, the spindle disappears but the spindle MTs that lie across the spindle equator remain to form the ...

(4) **Phragmoplast**. The phragmoplast consists of two sets of interdigitated MTs, possibly oppositely polarized, across the region where the cell plate will form. The MTs of the phragmoplast become progressively clumped together towards the periphery of the cell plate, which appears to push them aside as it forms. The cell plate forms by the fusion of vesicles, budded off from the dictyosomes, containing cell wall materials. The cell plate extends and fuses at its

outer edges with the existing side walls of the mother cell at exactly the position previously occupied by the PPB. Sometimes there is slightly greater deposition of cellulose on the parent walls at the future site of fusion of the cell plate, and sometimes a small anticipatory outgrowth of the parent cell wall. These observations suggest that at the position of the PPB there is something different about the plasma membrane or the cortical cytoplasm, but we have no evidence of what this difference might be.

After formation of the new cell wall, the phragmoplast disappears. In the daughter cells, the hoop reinforcement pattern of MTs ('shredded wheat') reforms from MTs apparently associated with the surface of the nucleus. These MTs seem to radiate from the nuclear envelope, which is thought to be an MT nucleating centre, i.e. the site where MTs form by crystallization of the tubulin protein. The MTs grow and extend very rapidly and wrap around the cell to form the helical arrays. These MTs then become linked by protein cross bridges to the plasma membrane.

The orientation of the cortical MTs (i.e. those just beneath the plasma membrane) matches the orientation of the cellulose microfibrils in the innermost layer of the cell wall on the other side of the plasma membrane (see Ch. 4, Fig. 4.4). In those few cells where there is not a match, it has been argued that this is because the MTs have very recently changed orientation but the new cellulose microfibrils corresponding to this orientation have not yet formed. It has also been argued that this is because the match is coincidental and not obligatory or functional. The general view is that the MTs in some way direct the orientation of the wall microfibrils, perhaps because the cellulose synthesizing complexes on the outer face of the plasma membrane can only move parallel to the MTs and cannot cross over those bands of the plasma membrane to which the MTs are attached on the inner surface.

The orientation of the innermost wall microfibrils is generally transverse to the axis of cell growth. How are they orientated? If they are orientated by the MTs, how are these orientated? Since MTs disrupted by drugs reform in their previous orientations, is it actually the MT organizing centres (MTOCs) that are orientated? If the cell plate forms in the plane of least shear, can the microfibrils or the MTs detect strain in the wall? These questions remain unanswered.

5.6.2 Microfilaments and intermediate filaments

These have been shown to be present in plant cells. The microfilament array seems often to be superimposed on the MT array. Microfilaments, as in animals, are composed of actin. Cytochalasin B, which disrupts microfilaments, also disrupts nuclear movement and cytoplasmic streaming. Microfilaments are thought to be the part of the cytoskeleton that positions and moves the nucleus. These various cytoskeletal elements should soon become better understood now that they can be observed by immunofluorescent techniques (Fig. 5.9).

5.7 THE CYTOSKELETON AND THE PLANE OF CELL DIVISION[8]

The PPB forms round the cell at the future division site and usually has the same transverse orientation as the cortical MTs. This being so, how then does the plane of division change? The only direct evidence comes from files of cells dividing longitudinally in the pea, in which the MTs and PPB were made visible by immunofluorescence. While most of the MTs were transverse, some seemed to have changed or be changing their orientation to become longitudinal. This would be consistent with observations from the *Azolla* root that the predominant transverse orientation of the MTs and wall microfibrils remains unaltered after longitudinal divisions. This implies that the plane of the PPB, and therefore the plane of division, does not depend on the orientation of the MTs or microfibrils in the parent cell but that MT and PPB orientation can be determined independently.

The orientation of the MTs in the cell after division seems to depend on cell shape in *Graptopetalum*. When the plane of division changes, the orientation of the MT arrays also changes only when the aspect ratio (length/width) of the cells is <0.7 (see also section 5.4.1). This may be because the MTs take up the energetically most favourable configuration and therefore tend to be parallel to the longest axis of the cell with the least number of corners, which would mean that they would tend normally to be looped round the diameter of a cylindrical cell rather than round its ends. Exceptions are tip-growing cells, where MTs are usually longitudinal rather than transverse to the growing cell axis.

The MTs may also become aligned to be normal to the predominant axis of growth. In the formation of grass (*Phleum*) stomata, the orientation of the MTs changes between cell divisions. In the young guard cell, the MTs are predominantly radial but change to predominantly longitudinal, and this precedes or accompanies a similar change in the orientation of the guard cell wall microfibrils. This change takes place at the time when the guard cells swell transiently in a transverse direction before they resume their longitudinal extension (Fig. 5.10). A similar shift also occurs in the adjacent subsidiary cells. Incidentally, the fact that the MTs were originally transverse implies that this was their orientation at the time of the division that formed the two guard cells and, therefore, at that

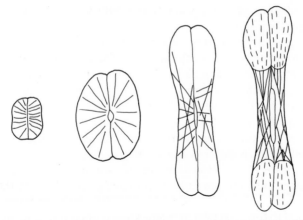

Figure 5.10 Changes in microtubule–microfibril orientation during stomatal formation in grasses. The microtubules change from transverse to longitudinal when the guard cells briefly expand sideways. (After Palevitz 1981)

longitudinal division the PPB must have been normal to the predominantly transverse MT array (as in the pea cells mentioned above). The opposite situation, in which a change in orientation of MTs seems to determine the direction of cell growth, is found in the tip cell of the fern protonema about to begin two-dimensional growth (see Ch. 6, section 6.3.2).

The PPB may not always be necessary in order for the cell plate to be orientated in a specific plane. There does not seem to be a PPB in tip-growing cells, such as the filaments of mosses, although divisions are transverse to the growing axis. It used to be thought that PPBs were absent from unorganized cells, such as callus or cell suspensions, but they have now been found here by immunofluorescent staining of the cytoskeleton. Using immunofluoresence microscopy, it has now been shown that in the shoot apex the orientation of the microtubules is apparently parallel to the plane of cell division in the tunica and the pith rib meristem.

The plane of division is important in plants because the cells remain fastened to each other and do not move, so the plane of division determines the spatial orientation to each other of daughter cells developing into different cell types, and the mechanical structure of the tissues that they form.

5.8 SUMMARY

(1) The plane of cell division is usually normal to the axis of growth, irrespective of cell shape, which therefore tends to be conserved. The cell plate tends to form so as to be of minimal area for halving

the cell volume, i.e. normal to the cell's long axis and orthogonal to existing walls.

(2) Change in shape may be associated with changes in rates of division (isotropic growth) or changes in planes of growth and division (anisotropic growth). Cell division can be inhibited by γ-irradiation or by inhibitors without affecting growth polarity. The plane of division is, therefore, a consequence and not a cause of the polarity of growth. An initial stage in the formation of leaves and lateral roots is a change in growth direction, which can occur in the absence of cell division.

(3) The plane of cell division can be orientated by pressure, either naturally, as in the cambium, or experimentally, as by the action of a piston on callus cells. The plane of division usually appears to be normal to the direction of major stress. In general, the new cell plate forms in the plane of least shear, which is usually normal to the main growth axis.

(4) Unequal divisions are forerunners of cell differentiation and produce an unequal distribution of cytoplasm between the daughter cells. In the formation of stomata, the division plane changes in successive divisions, but what causes this is not clear, although in the grass, *Phleum*, it may be transient changes in the direction of cell expansion.

(5) The cytoskeleton consists of microtubules, microfilaments, and intermediate filaments. Microtubule orientation is usually correlated with the orientation of new wall microfibrils, and both are related to cell shape. Microtubules form the preprophase band, which accurately predicts the site at which the new cell wall will form. They also form the mitotic spindle, which positions the chromosomes, and the phragmoplast, which is involved with the formation of the cell plate. The exact role and mechanism of the cytoskeleton in determining cell shape and the plane of division remains obscure.

(6) After microtubules have been disrupted temporarily by drugs, they can reform in their original orientation, suggesting that there are microtubule organizing centres (MTOCs) that control microtubule alignment.

(7) Immunofluorescent techniques now make it possible for more detailed studies to be made on how cytoskeleton structure and changes in structure are controlled and, in turn, how they may control the plane of cell division.

FURTHER READING

Furuya, M. 1984. Cell division patterns in multicellular plants. *Annual Review of Plant Physiology* **35**, 349–73.

Gunning, B. E. S. & S. M. Wick 1985. Preprophase bands, phragmoplasts, and spatial control of cytokinesis. *Journal of Cell Science Supplement* **2**, 157–79.

Lintilhac, P. M. 1974. Differentiation, organogenesis and the tectonics of cell wall orientation. III. Theoretical considerations of cell wall mechanics. *American Journal of Botany* **61**, 230–37. (Division in plane of least shear).

Lloyd, C. W. (ed.) 1982. *The cytoskeleton in plant growth and development*. London: Academic Press.

Lloyd, C. W. 1987. The plant cytoskeleton: the impact of fluorescence microscopy. *Annual Review of Plant Physiology* **38**, 119–39.

Lloyd, C. W., L. Clayton, P. J. Dawson, J. H. Doonan, J. S. Hulme, I. N. Roberts & B. Wells 1985. The cytoskeleton underlying side walls and cross walls in plants: molecules and macromolecular assemblies. *Journal of Cell Science Supplement* **2**, 143–55.

NOTES

1 Fuchs (1975), Jeune (1975), Sinnott (1944). (Cucurbits, *Myriophyllum* and *Tropaeolum*).

2 Jensen (1976), Stange (1983). (Differential division rates in cotton embryos and *Riella*)

3 Barlow (1969), Foard (1971), Foard *et al.* (1965), Haber & Foard (1963). (Inhibition of cell division)

4 Barlow (1987). (Aspect ratio)

5 Brown & Sax (1962), Lintilhac (1984), Lintilhac & Vesecky (1980). (Mechanical stress)

6 Palevitz (1981), Palevitz & Hepler (1974), Pickett-Heaps (1969). (Stomatal development)

7 Koshland *et al.* (1988). (Mitotic spindle *in vitro*)

8 Sakaguchi *et al.* (1988). (Microtubules and cell division planes in the shoot apex)

CHAPTER SIX

The cellular basis of polarity

6.1 DEVELOPMENT OF POLARITY IN *FUCUS* AND *PELVETIA* ZYGOTES[1]

To find out how a polar axis originates in the first place, it is necessary to start with a system lacking polarity and to follow what happens as polarity develops. We must start with a single unpolarized cell. This would be possible if we could use an undifferentiated apolar cell in suspension culture, but this would be difficult to recognize and manipulate. The only practical cells to use are the free-living zygotes of some algae, such as the seaweeds *Fucus* or *Pelvetia*.

The eggs are spherical and, apparently, completely symmetrical and apolar. They become polarized only when they are fertilized and the zygote settles on a substratum and germinates. The axis of polarity in the cell and the position of **rhizoid** emergence can be established by directional environmental signals, such as light, and so can be controlled experimentally (Table 6.1). As the zygote elongates and grows, it divides by the formation of a transverse cell wall, normal to the newly established long axis of the cell. The two cells formed are the rhizoid cell and the upper, thallus cell, which gives rise to the thallus of the plant.

The development of polarity in germinating *Fucus* zygotes was thought to be accompanied by the production of an electrical current. To test this, Jaffe lined up about 200 zygotes in seawater in a capillary tube and illuminated them from one end so that they all formed rhizoids in the same direction. He measured the voltage difference between the ends of the tube and found that a potential drop between rhizoid and thallus poles appeared when germination began. He calculated that the current required to generate this potential was equivalent to a flux of 100 pA per zygote, which is a very large ion flux for a plant cell. No potential drop appeared in a control tube in which the zygotes had germinated in diffuse light and so had grown out in all directions and were, therefore,

Table 6.1 Factors that can establish the polarity of algal zygotes. The effects of other eggs and algal thalli can be mimicked by the water in which eggs or thalli have recently been, indicating the existence of a soluble chemical factor released by the eggs or thalli into the water, which stimulates rhizoid emergence on the side of the zygote to which it is in greatest concentration. This chemical has been named rhizin; its chemical nature is not known. It was thought to be auxin but does not in fact seem to be identical. (After Jaffe 1969)

Factor	Position of rhizoid
unpolarized light	dark
polarized light	plane of electrical vector
sperm	point of entry
shape of egg made prolate	long axis
heat gradient	warmer end
centrifugal force (pH 8)	centrifugal
centrifugal force (pH 6)	centripetal
osmotic gradient	high water potential
voltage gradient (25 mV)	positive
voltage gradient (10 mV)	negative
K^+ gradient	high K^+
pH gradient	high H^+
dinitrophenol gradient	high DNP
indole acetic acid gradient	high IAA
flow	upstream
diffusion barrier (pH 8.5)	away
diffusion barrier (pH 6)	toward
another egg (pH 8; 1 egg diameter)	away
another egg (pH 6; 4 egg diameters)	toward
various algal thalli	toward

not in series. Jaffe later developed the vibrating probe electrode, which now allows measurements to be made on single cells (Box 6.1). In germinating *Fucus* zygotes, the current flowing through them results from cations entering at the tip of the rhizoid pole and leaving at the thallus pole. An important, but not necessarily major, component of the ion current has been shown to be carried by calcium ions.

The events during the development of polarity in the *Fucus* zygote are:

(1) The site of rhizoid formation is assigned by an environmental gradient, or, if this is absent, by the point of entry of the male gamete in fertilization. A polar axis develops.
(2) An ion current develops with cation (Ca^{2+}) influx (via calcium channels) at the presumptive rhizoid pole and cation (Ca^{2+}) efflux (via calcium pumps) at the presumptive thallus pole.
(3) Cytoplasmic vesicles, which contain cell wall precursors, accumulate at the rhizoid pole and this is seen as a clearing of the outer few micrometres of the cortical cytoplasm there.

This is a technique that allows the measurement of currents around cells without disturbing them. The principle of the vibrating probe is that the measuring electrode, which is a small (25 μm diameter) platinum sphere, is vibrated between two positions about 30 μm apart. By measuring the potential at each position with respect to a reference electrode situated some distance back, the potential difference between the two positions of the measuring electrode is obtained. The electrode is vibrated at about 200 Hz by a piezo-electric bender. The probe is placed normal to the surface of the cell being tested but does not touch it (Fig. 6.1). By taking measurements at various positions around the cell, a map of the current flow around the cell can be constructed and the current densities at the cell surface inferred. The probe has a voltage resolution of 1–2 nV and an inferred current density resolution of about 20 nA cm^{-2}.

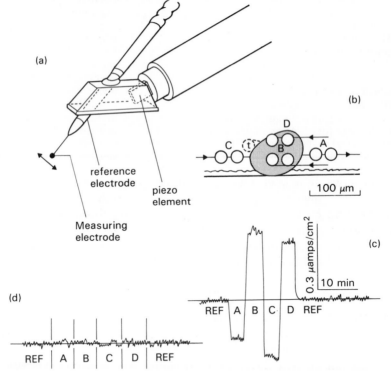

Figure 6.1 Measuring currents with the vibrating probe electrode. (a) The measuring electrode tip is placed about 50μm from the cell surface and vibrated between the two positions shown (o-o). The potential difference recorded on the trace is converted to a calculated value for the current. (b) The positions of the probe are shown for measurements on a *Lilium* pollen grain, which later produced a pollen tube at position t. (c) The current flux at positions A–D: downward deflections indicate flux left to right in (b); upward deflections indicate flux right to left. (d) Trace from a dead pollen grain. (After Weisenseel & Kicherer 1981)

(4) This site becomes stabilized as part of the process of the fixation of the rhizoid–thallus axis. The polar axis is now irreversibly established.

(5) New cell wall material, including fucoidin (a sulphated polysaccharide), is deposited at the rhizoid pole, which begins to bulge and extend by tip growth.

(6) Tip growth and cell wall synthesis continue to extend the rhizoid. The zygote divides into a rhizoid cell and a thallus cell by a division transverse to the axis.

The formation and establishment of the polar axis can be followed by seeing how the ion currents develop. Within 30 min of fertilization a transcellular current can be detected in *Pelvetia* zygotes. What is thought to happen is that influx of Ca^{2+} at the rhizoid pole leads to accumulation there because of the low mobility of Ca^{2+} within the cell and its tendency to bind strongly to proteins. A gradient of calcium concentration is therefore quickly established, the concentration at the rhizoid pole becoming about five times that at the thallus pole. Since the total transcellular flux of Ca^{2+} ions does not change during establishment of the ion current, this implies no change in the numbers of calcium channels and pumps but merely an alteration in their distribution in the plasma membrane of the cell. Calcium channels, allowing Ca^{2+} influx, become concentrated at the rhizoid pole and calcium pumps, causing Ca^{2+} efflux, at the thallus pole. The effect of the environmental polarizing factors is, therefore, to cause movement and localization of the calcium channels and pumps within the membrane, so that the random influx and efflux of calcium ions becomes concentrated into a transcellular current of about 1 μA cm^{-2}, which corresponds to 5 pmol cm^{-2} s^{-1} for a divalent cation such as calcium. It is not entirely clear that the calcium measured in these experiments had actually accumulated within the cytoplasm rather than remaining bound to the wall. New techniques, such as ratio imaging fluorescence microscopy, are now available to provide more reliable data.

6.1.1 *Fixation of the axis*[3]

The axis defined by the transcellular current does not become permanently stabilized until about 12 h after fertilization. This can be shown experimentally by using a unidirectional light stimulus to establish the axis initially and then testing the ability of the axis to become reorientated in response to a second light beam at 90° to the first. If the direction of light is altered more than 12 h after fertilization, then the direction of the axis can no longer be changed; the original direction has become fixed. Significantly, it is at this same time that the net movement of calcium

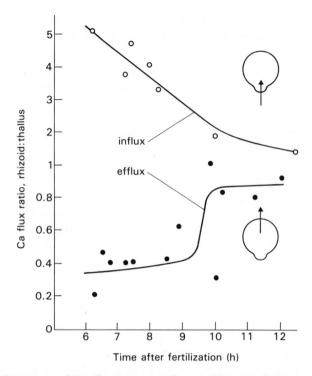

Figure 6.2 Changes in calcium flux during axis fixation. The ratio of calcium influx at the rhizoid : thallus pole decreases to unity by about 12 h after fertilization. Similarly, the calcium efflux ratio (rhizoid : thallus pole) rises to unity so that by 12 h, when the axis is fixed, there is no longer any difference in calcium influx or efflux between the poles of the zygote, and the internal calcium gradient disappears. (After Robinson & Jaffe 1975)

ceases and the flux ratio at the rhizoid and thallus poles becomes unity so that the unequal distribution of calcium within the cell, brought about by the transcellular current, finally disappears (Fig. 6.2).

What are the cellular events that cause axis establishment and fixation? Some clue can be gained from the effects of various inhibitors on the fixation process. Zygotes can be put in unidirectional light and, at the same time, be exposed to an inhibitor. If axis fixation is inhibited or delayed, as shown by removing the inhibitor and then testing with a second light beam 90° to the first, it will still be possible to alter the direction of the axis at a time when it has become fixed in control zygotes not treated with the inhibitor. Protein synthesis is not required for axis fixation, since cycloheximide (an inhibitor of protein synthesis) has no effect on axis fixation. Similarly, enlargement by water uptake is not essential, since high sucrose concentrations (which lower the **water potential** of the medium and prevent water uptake by the zygotes) do not affect axis fixation. Both of these treatments, however, delay rhizoid

104

growth. On the other hand, cytochalasin B (CB), which disrupts actin microfilaments in the cytoplasm, is an effective inhibitor of the process of axis fixation, but has no effect on the orientation of the axis once it has become fixed. Colchicine, which inhibits the formation and assembly of microtubules, has no effect on axis fixation. The fixation of the axis, therefore, seems to depend on the presence of the microfilaments in the cytoplasm. These may be involved in directing cytoplasmic vesicles to the site of rhizoid outgrowth, since actin becomes concentrated here. How the microfilaments themselves would become organized to do this is not known – perhaps they are orientated by the current flowing through the cell. Whether they would be able to direct vesicles by virtue of their possible contractile properties is also not known.

The presence of the cell wall may also be necessary for axis fixation. *Fucus* zygotes in a medium containing wall-digesting enzymes lost their cell wall. The fixation of the axis did not take place until they had been transferred back to artificial seawater (containing sucrose as a carbon source) and the wall had begun to reform. This may reflect a role for the wall as a mechanical framework for anchorage of the cytoskeleton by transmembrane connections.

6.1.2   Cell wall growth

At the site of rhizoid initiation there is an accumulation of mitochondria, dictyosomes, and associated vesicles in the cytoplasm, visible as a cleared region of the cell cortex, which can be reorientated, like the axis, by light. A sulphated polysaccharide, fucoidin, accumulates and is deposited in the cell wall specifically at the rhizoid site. Fucoidin localization is inhibited by CB, which does not, however, interfere with its sulphation. Cytochalasin B, therefore, inhibits only the localization of the fucoidin, not its synthesis. Since CB also inhibits axis fixation, the inference is that the process of localization of fucoidin deposition is an essential part of axis fixation. However, if sulphation is inhibited by growing zygotes in sulphate-free seawater, the amount of fucoidin in the cell as a whole remains unchanged, as shown by binding assays, but it does not accumulate at the rhizoid pole. Although the zygotes do not adhere to the substratum, the rhizoid still forms and so axis formation continues. The presence of sulphated fucoidin, therefore, is not essential for rhizoid formation. Since the unsulphated fucoidin does not become localized, it seems that it is not the nature of the substance deposited or the localization of its deposition but the establishment of an intracellular directional transport system that is the essential process in axis fixation. The actual localization of substances such as fucoidin is consequent on axis fixation, but it is not necessary for it to take place.

How then could such a directional transport process work? Two possible mechanisms have been suggested. The first is self-electro-phoresis. The establishment of the transcellular current means that initially in the cell there is a gradient of electrical potential. This results from the ionic gradient, created by the accumulation of positive charges at the rhizoid pole and negative charges at the thallus pole, which gradually declines and disappears when the axis becomes fixed (Fig. 6.2). Negatively charged particles within the cell will, therefore, tend to move by electrophoresis towards the rhizoid pole. Many proteins carry a net negative charge and so this could possibily account for the movement of proteins and vesicles in the cytoplasm to the rhizoid pole. It might also suggest that sulphation of fucoidin to give it a net negative charge is required for it to become concentrated and localized at the rhizoid pole. The sulphated fucoidin has been shown to be within cytoplasmic vesicles, which bear a greater net negative charge when the fucoidin in them is sulphated.

The second mechanism proposed is based on contractile filaments. The need for microfilaments for axis fixation suggests that these may be essential for the directed movement of vesicles to the rhizoid site. How this might be achieved is, at present, unknown. The establishment of polarity may, however, first require the organization of oriented cytoskeletal structure in the cell.

6.2   POLARITY IN OTHER CELLS – ION CURRENTS

To what extent is the establishment of polarity in algal zygotes typical of other systems? In all plant cells and organs examined so far that show polarity, ion currents both precede and accompany growth. The current enters at the root or rhizoid pole and exits at the thallus pole or in the proximal part of the organ (Fig. 6.3). The inward current is typically accompanied by a calcium influx. When the plant poles are classified according to their function (Table 6.2), it is seen that the thallus pole corresponds to the photosynthetic pole and this carries the outwardly directed current. But even here, where the current may be carried by a net efflux of cations or protons, there may still be a small calcium *influx*. It is not yet clear whether this always precedes growth at either pole of the plant axis.

While it is not difficult to understand the existence of currents in the medium surrounding an absorptive organ, such as a root or an algal zygote immersed in water, it is not so obvious that an aerial photosynthe-sizing cell, filament, or organ can carry a current exiting at the growing pole and flowing in at the more basal regions. Measurements on fern protonemata have shown that the current density around a filament immersed in bathing solution is not only lower than that for absorptive

106

Table 6.2 Ion currents in relation to function in tip-growing structures. (After Cooke & Racusen 1986)

Tip-growing structure	Organism	Function	Net direction of positive tip currents
brown algal rhizoid	*Pelvetia* *Fucus*	absorption	inward
red algal rhizoid	*Griffithsia*	absorption	inward
fungal hyphae	7 genera	absorption	inward
fungal rhizoid	*Blastocladiella*	absorption	inward
angiosperm pollen tube	*Lilium*	absorption	inward
angiosperm root hair	*Hordeum* *Lepidium*	absorption	inward
brown algal thallus initial	*Pelvetia* *Fucus*	photosynthesis	outward
yellow-green algal filament	*Vaucheria*	photosynthesis	outward then inward
moss protonema	*Funaria*	photosynthesis	?
fern protonema	*Onoclea*	photosynthesis	outward

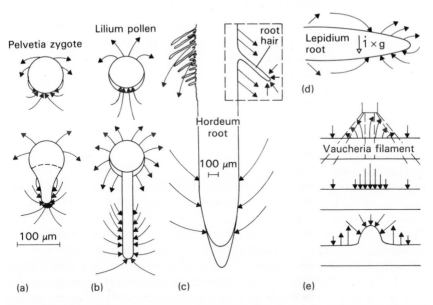

Figure 6.3 Natural ion currents generated by growing plant cells and organs. Current enters at the absorptive pole and exits proximal to the growing tip or at the photosynthetic pole. When roots are gravitropically stimulated by being turned on their sides (d), an outward current develops on the upper side of the tip within 3 min. (Weisenseel & Kicherer 1981; Behrens *et al.* 1982)

structures, such as algal zygotes and lily pollen tubes, but is not sufficient to account for the measured potential difference. This may be because, in this case, the major current pathway is not in the external medium but in the epidermal cell walls, outside the plasma membrane but inside the relatively impermeable cuticle. If this is so, then it raises the possibility that ion currents may also be a feature of the growth of completely aerial structures, such as the shoot apex of land plants, on which measurements have not yet been made. In view of the apparent importance of calcium in the regulation of cellular activity (see Ch. 12, section 12.7), calcium influx, which always seems to occur at the growing pole, may yet prove to be the key essential feature in the establishment of a polar axis.

To generate an electrical field around a cell, the distribution or activity of ion channels and ion pumps in the membrane must alter so that, instead of an even distribution of both over the cell surface, active channels become concentrated at one site and pumps at another. This can be triggered, as we have seen, by a variety of external stimuli, but how these stimuli are transduced and bring about the membrane changes is at present unknown. Although algal zygotes, pollen tubes, and germinating *Equisetum* and *Funaria* spores grow in alignment with an imposed electric field, the growing point is often directed toward the cathode but can be directed to either anode or cathode. This would not be expected if self-electrophoresis were involved in specifying at which end a particular pole is located.

6.3   CHANGING THE POLARITY OF GROWTH

6.3.1   *Branching in* Vaucheria

Ion currents are involved not only in the establishment of a polar axis in non-polar systems but also in the establishment of a new axis on an existing one. In *Vaucheria* (a filamentous alga), a branch could be caused to form by shining a spot of blue light on the side of the filament (Fig. 6.4). The formation of a branch was always preceded by an outward current, although during the actual branch formation the current changed to an inward one. The outward current was carried mainly by protons. If ion currents are truly causal to the formation of a new polar axis, it should be possible to cause branching directly by the imposition of an electrical field on cells; this does not yet seem to have been achieved.

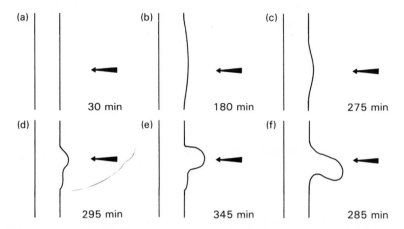

(a)   30 min   (b)   180 min   (c)   275 min

(d)   295 min   (e)   345 min   (f)   285 min

Figure 6.4 Branch formation in *Vaucheria*. A spot of blue light shone on the side of the algal filament induces an outward current, which is then replaced at the site of branch formation by an inward current (see Fig. 6.3e). After 3–5 h, a branch becomes visible. (After Weisenseel & Kicherer 1981)

6.3.2 Transition to two-dimensional growth in fern gametophytes[4]

Another instance of branching associated with the occurrence of an electrical current is the transition from unidirectional to bidirectional growth in the fern protonema, which grows on to produce the **gameto-phyte** prothallus. On germination, the spore forms a filamentous proto-nema of green, chlorenchymatous cells bearing a few rhizoids. Cell number in the filament increases by transverse divisions in the apical cell. Although some of the basal cells divide to produce rhizoids, the main filament continues to grow in length until the plane of division changes to longitudinal. In most species studied, this takes place in the apical cell itself and is the beginning of the two-dimensional (2-D) growth of the gametophyte (Fig. 6.5).

The transition to 2-D growth is triggered by blue (or white) light. In darkness or in red light, the apical cell divides transversely and the protonema remains filamentous. Transfer to blue or white light (at the same fluence rate) causes longitudinal division and the transition to 2-D growth. If growth is in blue or white light from the start, the transition is not immediately on germination but only after a characteristic number of transverse divisions, usually three or four.

Within 10 min of the transfer to blue light, the electrical and ionic characteristics of the apical cell begin to change. The extracellular electrical field declines and a small outward current becomes measurable just below the tip of the cell. This is the region that expands before the plane of division changes, although this may not occur until two cell

109

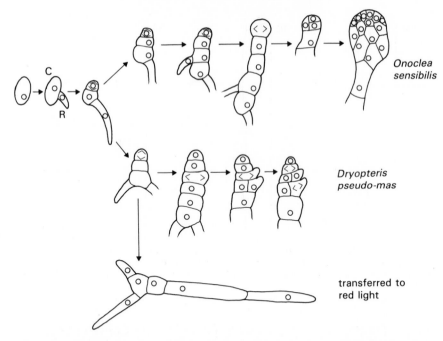

Onoclea
sensibilis

Dryopteris
pseudo-mas

transferred to
red light

Figure 6.5 Transition to two-dimensional growth in fern protonemata. The first division of the spore gives a rhizoid (R) and a chlorocyte (C), which divides transversely. The apical cell then continues to divide transversely to form a filament. In white or blue light, after it has formed 4–5 cells, the apical cell (*Onoclea*) or a subapical cell (*Dryopteris pseudo-mas*) divides longitudinally. Division then continues in various planes to form the prothallus. In red light the cells are elongated, the apical cell divides only transversely, a long filament forms, and there is no transition to two-dimensional growth. (After Dyer & King 1979; Miller 1980)

divisions later. The orientation of cytoplasmic microtubules also becomes more random in the apical cell, which is consistent with its more isotropic growth as it rounds up and becomes broader (Fig. 6.6). We do not know whether the change in microtubule orientation determines the change in cell shape or results from a change in cell shape brought about by localized changes in wall plasticity.

How could light quality alter cell shape and so alter division and growth planes? The primary effect of light is apparently to cause a redistribution of ion channels and pumps in the apical cell and to alter the properties of the growing cell walls, especially in the apical cell. Since the protonema shows tip growth, there is presumably a gradient of wall plasticity from a maximum at the tip of the domed apical cell to a minimum at its base (see Ch. 4, Fig.4.1a). Plasticity would have to increase lower down the sides of the apical cell for the cell to tend to bulge. The redistribution of ion pumps does cause a greater proton efflux at the sides of the apical cell and it may be this that causes wall loosening here and thus swelling (see Ch. 8, Box 8.2).

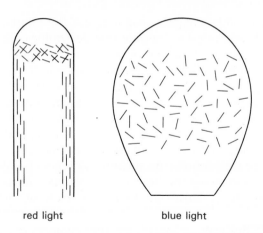

red light   blue light

Figure 6.6 Changes in microtubule orientation in the fern apical cell. In the *Dryopteris filix-mas* apical cell, before the change to longitudinal division that leads to two-dimensional growth, the microtubules are orientated axially in the basal part of the cell but randomly just below the growing tip, from which they are absent. On transfer to blue light, the microtubules become randomly orientated throughout the cell. This is associated with swelling of the cell and the change in division orientation from transverse to longitudinal. (After Stetler & Demaggio 1972)

After the apical cell swells the next division is longitudinal, instead of transverse, and so the 2-D growth pattern is initiated. Why does the plane of division change? When the areas of the new cell wall were calculated from the lengths and widths of apical cells at the position of the mitotic nucleus, in 46 out of 49 cells measured the plane of the division, whether transverse or longitudinal, was the one that gave the wall of least area. These new cell walls were also in the plane of least shear (see Ch. 5, section 5.4.4), normal to the principal direction of stress as calculated from cell shape. This is consistent with the hypothesis that the stress pattern of the cell affects the orientation of the mitotic spindle. The principal direction of stress becoming transverse could result simply from the change in the shape of the cell, as a result of the change in wall microfibril orientation, whether or not this was accompanied by a change in its relative longitudinal and transverse growth rates.

Observations of protonemata had previously suggested that the change in apical cell shape might be the consequence of changes in the relative rates of cell elongation and cell division. If the rate of division becomes higher relative to the rate of elongation, as it does in blue light, then the cells will become shorter but not necessarily fatter. What happens in *Dryopteris pseudo-mas* is relevant here, since the transition to 2-D growth in this species occurs not in the apical cell but usually by a longitudinal division in the third cell behind the apex (Fig. 6.5). When this cell is longer than normal the division tends to be transverse instead of longitudinal. This suggests that it is not the relationship between elongation rate and cell division rate that determines cell shape but the relative longitudinal and transverse cellular growth rates. This is shown by the effect of light quality on protonemal and cell shape. In red light the cells are long and thin. In blue light the cells are shorter and fatter. However, irrespective of light quality, the protonemal volume can remain the same. This means that what is really under the control of light is the cross sectional area of the cells and this will depend on the relative transverse and longitudinal growth rates in the filament. Since we would expect this to be a function of the anisotropy of the cell wall structure (see Ch. 4, section 4.2), what the light is presumably controlling (as in the apical cell) are the relative amounts of transverse and longitudinal microfibrils in the wall and the plasticity of the side walls of the filament.

Although in this way we can explain the transition to 2-D growth, it is not clear why the prothallus continues to grow as an essentially two-dimensional structure and does not become three-dimensional to any marked extent. There must be further controls that dictate the subsequent polarity after the transition to 2-D growth. This is similar to the problem of the dorsiventral leaf, which grows as a lamina with only limited growth in thickness.

In fern protonemata, although phytochrome is apparently a photoreceptor (because the effects of red light on cell cross sectional area are reversed by far-red light), a blue-absorbing pigment (cryptochrome) must also be involved because the relative effectiveness of light at 420 and 660 nm does not correspond with what would be expected of phytochrome alone, and red and far-red light alone do not induce the normal transition to 2-D growth. The photoreceptor system is located in the apical 50 μm or so of the apical cell, as shown by microbeam irradiation. In *Dryopteris pseudo-mas*, the first longitudinal division is in a subapical cell, implying the transmission of some message from the apical to the subapical cell.

6.4   POLARITY IN UNEQUAL DIVISIONS

Unequal cell divisions are a feature of plant development, especially where the daughter cells differentiate from each other – so much so that unequal division is believed to be a necessity for subsequent differentiation. The first division of a zygote is nearly always asymmetrical and this is the first of many such unequal divisions during development. The closest study of unequal division has been made in small plants, such as *Fucus* germlings and fern gametophytes, for the same reason that they have been used for other studies: being free-living, they are easily accessible and amenable to experimentation.

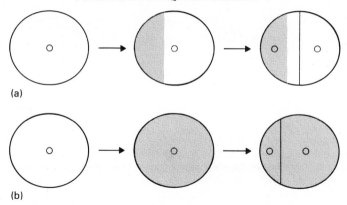

Figure 6.7 Possible types of unequal division: (a) a symmetrical division may divide an asymmetrical cytoplasm; (b) a symmetrical cytoplasm may be divided into larger and smaller cells by an unequal division.

There are basically two types of unequal division:

(1) Cytokinesis equally divides a cell in which the cytoplasm is asymmetrically distributed and so the daughter cells receive unequal amounts.
(2) Unequal cytokinesis divides a more or less symmetrical cell into a smaller and a larger daughter cell, each therefore containing different amounts of cytoplasm or cytoplasmic components, e.g. vacuoles (Fig. 6.7). Unequal cytokinesis could also, presumably, cause a more or less equal division of asymmetrically distributed cytoplasm.

The first type is typical of what happens in the first division of algal and higher plant zygotes: the more or less symmetrical division divides an already polarized cytoplasm. The second type, giving daughter cells of very different sizes, is typical of the first division of the germination of fern spores, of pollen grain mitosis in higher plants, and of the divisions involved in the differentiation of stomata and in the formation of root hairs. This seems to be the sort of division usually involved in the initiation of differentiation in situations as diverse as the formation of the sieve tube and its associated companion cell in the vascular plant, and the division of the apical cell in **bryophytes** and pteridophytes.

6.4.1 Onoclea *fern spores*[5]

6.4.1.1 Events of unequal division
In the tetrad of four spores that results from a meiotic division, each spore is symmetrical in shape, being an oblate spheroid with a flattened proximal face where it was in contact with the other three spores. Before

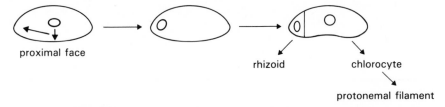

proximal face

rhizoid   chlorocyte

protonemal filament

Figure 6.8 Nucleus migration in *Onoclea* fern spore germination. The nucleus migrates first to the proximal face, which is the site of heavy-metal binding (see Fig. 6.11), and then to one end of the cell where it divides. The smaller cell forms a rhizoid and the larger a chlorocyte. It is not known whether this division corresponds to (b) in Fig. 6.7 or to a combination of (a) and (b). (After Miller 1985)

mitosis, the nucleus migrates to the centre of the proximal face and then towards one end (Fig. 6.8). Cytokinesis then produces a small lens-shaped cell, which forms the rhizoid, and a larger cell, which becomes a chlorocyte. Normally, the rhizoid cell never divides again, and its nucleus does not synthesize any more DNA. The chlorocyte goes on to divide to form the filamentous protonema.

6.4.1.2 Necessity of unequal division for differentiation
Several experimental treatments can result in equal instead of the usual unequal division. These treatments prevent the nucleus from migrating in the normal manner. If, for example, the spores are centrifuged to displace the nucleus after it has completed its migration but before division, then the nucleus does not migrate back to its proper position and the resulting division is equal (symmetrical). Those spores in which division is unequal differentiate a rhizoid, but those in which it is equal do not.

As a result of centrifugation, the cell contents and organelles are drastically rearranged. Despite the stratification of the cell contents, if the division itself is unequal then a rhizoid differentiates. But if the division is equal, irrespective of the fact that the cell contents are asymmetrically distributed, the rhizoid does not differentiate. This shows that in this cell it is the unequal division itself that is essential for differentiation and not the asymmetrical partitioning of the cytoplasmic contents. It seems to be simply their quantity that matters, irrespective of their composition.

Other treatments that prevent nuclear migration and subsequent rhizoid differentiation include the effects of alcohols or other lipophilic solvents, and substances such as colchicine, which disrupt or complex with microtubules. The alcohols are effective in direct proportion to their lipid solubility, which suggests that they are acting by disrupting some lipophilic site in the spore, perhaps in the membrane, that is essential for nuclear migration. The effect of the drugs that disrupt the cytoskeleton is probably because the cytoskeletal elements move and position the nucleus. This has been studied in more detail in **tip growth** (see section 6.5.3). Differentiation is also prevented by caffeine, which prevents the formation of the cell plate between the daughter cells so that a binucleate cell is formed. Differentiation does not take place unless there is subsequently an unequal division following a further mitosis.

6.4.1.3 Localized ion accumulation in relation to nuclear migration

The outer thick wall of the spore is the exine, and outside this is a loose, thin, brown spore coat, the perine. When the perine is present, spores can germinate and the rhizoid can grow out in distilled water. If the perine is removed, by a few minutes' treatment with sodium hypochlorite, then the spores need to be supplied with Ca^{2+}, Mg^{2+} or Mn^{2+} ions in order to germinate normally. Fragmented material from isolated perines can substitute for metal ions, suggesting that the perine normally acts as a store of the necessary ions and can supply them to the spore. When spores are stained to show the presence of heavy metals, the stainable material is found to be concentrated on the proximal face of the spore and can be removed by hypochlorite. If the spores are first treated with hypochlorite, which presumably removes the native heavy metals, then they can absorb ions supplied from the culture medium and these accumulate preferentially on the proximal face of the exine. The ability of the spore face to absorb heavy metals changes during germination (Fig. 6.9). Absorption reaches a peak after about 2 h, decreases a few hours later and then again increases. If the ability of the exine to sequester exogenously supplied ions reflects its ability to sequester native ions, then this might suggest that during germination there is a movement of ions to and fro between the spore and the exine. The natural ion of most importance is thought to be calcium. The implication would be that the activity of calcium pumps and channels on the proximal face of the spore changes in a systematic fashion during germination. The particular interest is that the proximal face of the spore, where the ions are concentrated, is also the face of the spore towards which the nucleus first moves during the migrations leading up to the asymmetrical cell division.

Can the two phenomena be linked? The migration of the nucleus does not begin until the ions (presumably calcium) become concentrated in the exine for the second time (Fig. 6.9). When germinating spores were treated with 8 mM colchicine (which depolymerizes microtubules), nuclear movement, mitosis, and cell division were prevented but the spores continued to grow. After 42 h the spores were removed from the colchicine and placed on agar, so that they were fixed in orientation. Twelve hours after removal from colchicine, visibly polarized cell activity was resumed but the nucleus remained central and the subsequent cell division was symmetrical. These spores were allowed to continue devel-

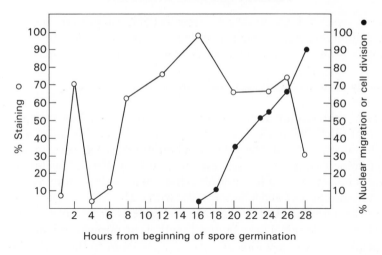

Figure 6.9 Changes in metal-binding to the proximal face of the fern spore. Metal-binding was measured by sulphide–silver staining of germinating spores of *Onoclea sensibilis*. The second peak of metal-binding is followed by nuclear migration. (After Robinson *et al.* 1984)

opment in plane-polarized red light, which does not affect this first division; it therefore seems to be orientated (although symmetrical) by an inherent polarity in the spore. However, the next division was orientated by the polarized light so that nuclear migration and spindle orientation tended to be parallel to the plane of polarization and the division plane was therefore normal to the plane of polarization. What is particularly interesting is that the stainable patch due to heavy metals on the face of the spore was also at the position to which the nucleus would be expected to migrate under the influence of the polarized light. The implication is that the migration of the nucleus may, therefore, be in some way controlled by the position of ion pumps and channels and that it is the position of these that is altered or affected by the polarized light.

Other unequal divisions occur in fern protonemal development during the initiation of **antheridia**, rhizoids, and prothallus. In all these, before nuclear migration, Ca^{2+} accumulates in the cytoplasm and wall at the site of the future division. It has been suggested that this could cause a decrease in Ca^{2+} concentration in the region of the nucleus and, in ways not yet clear, play a key role in regulating cytoskeletal constituents that control the migration of the nucleus. Again, it seems probable that it is calcium ion pumps and channels that are important and that ion currents and changes in their position and intensity on the cell surface are precursors to orientated events in the cell, in this case leading to unequal division.

6.5   POLARITY IN TIP GROWTH[6]

6.5.1   *Nuclear migration in* Adiantum *fern protonemata*

In growing fern protonemata, growth is principally at the tip of the filament. The nucleus remains about 60 μm behind the tip, and to maintain this position it therefore has to continually migrate forward at the same rate as the tip grows. When the cell is about to divide, tip growth temporarily stops, the nucleus moves a short distance backwards (Fig. 6.10), and mitosis and cytokinesis follow. To find out how the nucleus is positioned and, therefore, what may also be the mechanism of its movement, experiments have been done that involve displacing the nucleus backward (**basipetally**) by centrifugation at specific times during the cell cycle. In order to get cells in a specific part of the cell cycle the technique is to grow the filaments in red light, in which cell division will not occur, and then to transfer them to darkness, which stimulates the cells to go on to divide (Fig. 6.10). A centrifugal force of 110 g for 15 min is insufficient to displace the nucleus in a cell that is some time before mitosis, i.e. early in the G_1 part of the cell cycle (Fig. 6.11). At the late G_1 phase, some 20 h after transfer to darkness, the nuclei in more than half the filaments are displaced, but in G_2 and mitosis the nucleus again becomes more difficult to displace. This implies that the nucleus is more firmly held in place just before, during, and just after division.

The effects of various inhibitors give some idea of which subcellular structures may be involved in holding the nucleus in position. Thirty-eight hours or so after the transfer to darkness, when the nuclei are about to divide and are mostly not displaced by centrifugation in untreated cells (Fig. 6.11), treatment with colchicine (which disrupts microtubules) results in displacement of more nuclei in centrifuged cells. This implies that microtubules are involved in holding the nucleus in position. Cytochalasin B, on the other hand, which disrupts microfilaments, does not affect the ability of the nucleus to resist displacement, so microfilaments are presumably not involved in nuclear positioning in these cells.

The distribution of microtubules in the cell tip changes during the cell cycle in a way that is consistent with these conclusions. The microtubules in the basal part of the filament are predominantly longitudinal (axial), but just at the base of the rounded dome at the tip of the filament they are arranged randomly or circumferentially round the filament. At the extreme growing tip, there are few or no microtubules. The circumferentially arranged microtubules move from the tip region to the position of the nucleus, i.e. they are displaced about 90 μm in 16 h (Fig. 6.12). Over this same period (20–36 h after transfer to darkness) the filament grows by only about 20 μm, as it comes to a stop before cell division, so the

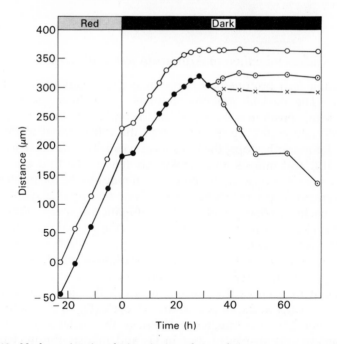

Figure 6.10 Nuclear migration during tip growth in a fern protonema. Apical cells of *Adiantum* protonemata were made to divide by transferring them from red light to darkness. The growth of the apical tip (-○-) slowed down and stopped. The nucleus (-●-) migrated to keep about 60 μm behind the tip. Just before division, it moved backwards about 20 μm. After division and the formation of the new cell wall (-x-), one daughter nucleus migrated forward, keeping 60 μm behind the tip and becoming the nucleus of the tip cell, and the other migrated backward about 150 μm to its position in the subapical cell. (Wada *et al.* 1980)

displacement of the microtubules is a true basipetal displacement; they do not simply remain stationary as the filament grows on beyond them for a while. The arrival of the microtubules in the region of the nucleus coincides with the increasing difficulty in displacing the nucleus by centrifugation. The simplest interpretation is that these circumferential microtubules are in some way involved in anchoring the nucleus prior to division. They disappear at prophase and therefore are a preprophase band (PPB) (see Ch. 5, section 5.6.1). The apparent existence of an invagination, containing microtubules, at the leading edge of the nucleus well before division, when it is maintaining its position relative to the filament tip by continuous migration forwards, has led to the speculation that microtubules are also in some way involved in nuclear positioning during this phase too, but in this case it would have to be the longitudinal microtubules that are involved. Microfilaments may be more concerned with the movement of cytoplasmic organelles, since treatment with cytochalasin B disrupts cytoplasmic streaming, which is also involved in

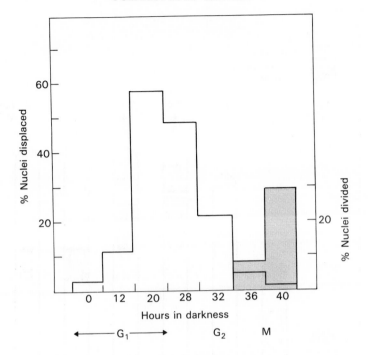

Figure 6.11 Effect of centrifugation on nuclear positioning in *Adiantum* tip growth. As the nuclei progress through the G_1 phase of the cell cycle, they become more easily displaced (clear graph) by a basal centrifugation of 110 g for 15 min, but they become more firmly held in place as division (shaded graph) is approached. (After Mineyuki & Furuya 1986)

the events leading up to cell division, as shown by the changing distribution of the organelles as the cell approaches division.

Sievers and Schnepf make the interesting suggestion that the prepro-phase band (PPB) may be involved in the positioning of the nucleus of cells that have no mechanism for transporting the nucleus in interphase, which would perhaps explain the almost ubiquitous occurrence of the PPB in plant cells but its apparent absence from cells that show nuclear migration as a normal feature of growth. However, this does not explain its presence in fern protonemata or in cells such as stomatal mother cells in which nuclear movement is commonly observed, unless nucleus positioning in some cells can be by microtubules and in others by microfilaments.

The basis of polarity in tip growth may lie in the distribution of ion pumps and channels, which could affect the distribution of calcium in the cell and, in turn, perhaps control and modulate the distribution of organelles and the position of the nucleus. Gradients of decreasing Ca^{2+} concentration with distance from the tip have been shown in tip-growing cells (Fig. 6.13).

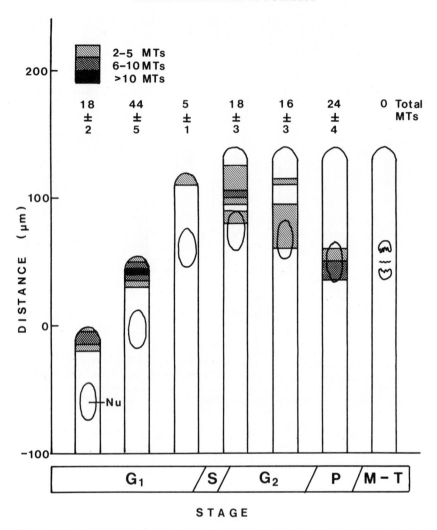

Figure 6.12 Movement of circumferentially arranged microtubules in the apical cell of a fern protonema. These microtubules in the *Adiantum* filament move backward (as the nucleus, shown in outline, moves backward), forming a preprophase band at the position where the cell plate will form. These microtubules disappear during mitosis. (After Wada *et al.* 1980; figure kindly supplied by Professor Y. Mineyuki)

6.5.2 The role of the nucleus in polarity in tip-growing cells

One of the manifestations of polarity is the position of the nucleus but this, in turn, can affect the polarity of the cell by determining the direction of the polar axis. In the experiments with fern protonemata, it was found that when the nucleus was displaced by a gentle centrifugation it could migrate back to its 'proper' position relatively rapidly.

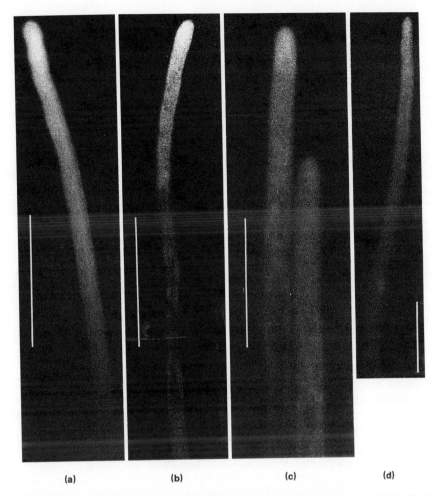

(a)   (b)   (c)   (d)

Figure 6.13 Calcium gradients in tip-growing cells. When chlorotetracycline (CTC) binds to calcium, it fluoresces. Tip-growing cells treated with CTC fluoresce, showing a gradient of calcium decreasing from the tip. (a) *Lilium* pollen tube. (b) *Lepidium* root hair. (c) Moss (*Funaria*) caulonema. (d) Fungal (*Achlya*) hypha. Bar: 100 μm. (Reiss & Herth 1978, 1979; photographs kindly supplied by Professor W. Herth)

After a strong centrifugation the nucleus was not always able to do so but remained displaced basally in the filament. When this happened, cell division and cytokinesis were also displaced basally and the filament just basal to the new cell plate bulged out at one side to form a branch. The basal position of the nucleus was, therefore, associated with the formation of a new axis at the side of the filament. In the moss *Funaria*, the new axis may even be directed backward toward the base of the protonema.

6.5.3   Polarity of structure in tip-growing cells

The internal organization is similar for all rapidly growing tip-growing cells, whether they are moss caulonema, pollen tubes, or fungal hyphae. (Slower growing cells with tip growth, such as fern protonemata [<10µm h^{-1}], do not show this internal organization.) At the extreme tip of the filament, the cytoplasm contains many dictyosome vesicles, which probably transport some polysaccharide wall precursors to the site of wall growth at the tip where they discharge their contents into the wall by exocytosis. The wall grows only in the curved region at the tip, with a maximum rate at the tip itself. As long as the rate of wall extension is equalled by the rate of supply of membrane material from the fusing vesicles, then the plasma membrane will be synthesized at a rate that depends on the rate of supply of membrane material. In pollen tubes, this is what seems to happen but, in *Funaria* caulonemata, the incorporation of material from the dictyosome vesicles has been calculated to be 5–10 times higher than can be accounted for by the growth rate of the plasma membrane. This implies that there is a constant recycling of plasma membrane material, the rate of membrane growth being the net result of the rates of accumulation and recycling back to the cytoplasm of membrane material.

This extreme apical region can be seen in the light microscope as the apical body. Behind this is the subapical zone, several micrometres in length, which, in addition to dictyosomes, contains mitochondria, endoplasmic reticulum (ER), and (except in fungi) plastids (Fig. 6.14). More basal still is the region where the large central vacuole is found and where microtubules and microfilaments are more in evidence. Even more basal (60 µm or more) is the nucleus. When this organization is temporarily disrupted by inhibiting the growth of the cell, or by centrifugation, it can eventually reform, and if the nucleus has been displaced by centrifugation it can migrate back to its normal position. This implies some basic structures that determine the organization, but which themselves are not altered by the experimental treatments that disrupt growth. Alternatively, the components of the apical system may have the intrinsic property of self-organization, but how they would do this is difficult to imagine. Whether the calcium gradient determines the internal structure or is a consequence of it is not yet known.

6.6   POLARITY OF GROWTH IN INDIVIDUAL, ISOLATED CELLS

The axial growth of cells or filaments comes about because growth is localized to the tip. There are three associated problems. How is the direction of the axis established in the first place? How is the direction of

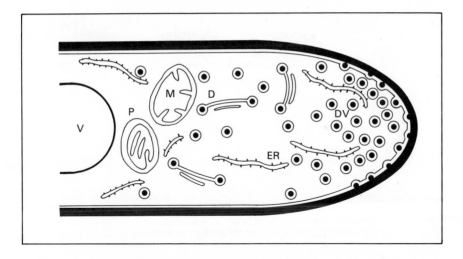

Figure 6.14 General organization of a cell showing tip growth. The curved tip region, where the wall is synthesized, is rich in dictyosome vesicles (DV), which add wall material by exocytosis. Just behind the tip there are dictyosomes (D) and endoplasmic reticulum (ER). Further back are mitochondria (M) and plastids (P) (except in fungi), then the vacuole (V) and behind this the nucleus. (After Sievers & Schepf 1981)

the axis changed or a new secondary axis established? What is the role of the nucleus? The first problem has been considered in the *Fucus* zygote, and the second in the changeover to 2-D growth in the fern protonema and in branching in *Vaucheria*. The second and third problems are particularly addressed when we consider the generation of cell form in the unicellular alga *Micrasterias*.

6.6.1   Polarity and the generation of form in Micrasterias[7]

Micrasterias is a **desmid**, which is a curious type of unicellular, fresh-water alga in which each organism consists of two semicells joined by an isthmus in which the nucleus resides (Fig. 6.15). When the cell divides, the nucleus divides, one daughter nucleus moves into each semicell, a septum is formed across the isthmus (growing in from the sides, not formed as a cell plate), and the cell wall at the septum bulges, each of the two abutting bulges growing into a new semicell with the same shape and pattern of lobes as the mother cell. There is a single chloroplast in each semicell. When a daughter semicell forms, after cell division, the chloroplast from the mother semicell bulges into the young, growing daughter semicell, and when this has attained full size the chloroplast divides at the isthmus to form a chloroplast in each of the semicells.

123

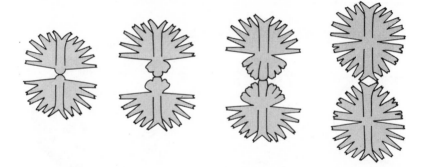

Figure 6.15 Structure and growth of *Micrasterias*. After nuclear division, a septum forms across the isthmus, thus separating the semicells. The wall of each semicell at the isthmus bulges and grows to form the three major lobes, which then grow and subdivide to form the minor lobes until a new semicell matching the parent one is complete. There are about 4 h between each of the stages shown here. (Kallio & Lehtonen 1981)

Problems posed by *Micrasterias* are:

(1)   How is the very specific cell shape of the new semicell achieved?
(2)   What is the role of the nucleus in the determination of cell shape?

Micrasterias was chosen for research because it is relatively large for a single cell (about 200 μm in diameter), is flattened, grows and develops at a convenient rate (the whole cell cycle takes about 3–5 days), has a distinctive cell shape, and is readily grown in culture. Most importantly, the growth and cell shape can be modified experimentally.

In order for growth of the new semicell to occur, there must be sufficient turgor. The osmotic pressure within the cell declines steadily as the cell grows but if the cell is placed in an osmoticum that is strong enough, cell growth is prevented. The cell wall still continues to grow but, because it cannot extend, cell wall material accumulates. The interesting thing is that it does so locally, at positions that correspond to the tips of the lobes (Fig. 6.16). It does not accumulate at the positions corresponding to the furrows. When the cell is restored to a solution of much lower osmotic pressure, which then allows growth to resume, the positions where wall material has accumulated grow so that the wall there thins out and the cell takes up its normal shape. This is an example of 'stored' growth. The implication is that the shape of the cell is determined by the differential synthesis of wall material at different positions on the cell surface. This means that there is some sort of pattern present on the surface of the young semicell. How is this pattern formed and what does it consist of?

The pattern is determined immediately by the cytoplasm in the growing

124

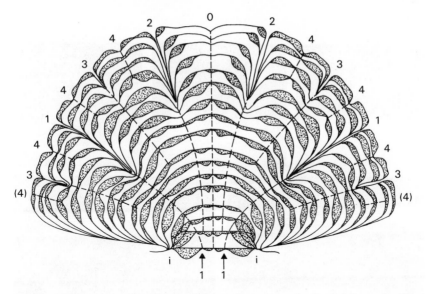

Figure 6.16 The pattern of wall accumulation under turgor reduction in *Micrasterias*. When the cell is placed in an osmoticum that prevents expansion, cell wall accumulation continues (stippled areas) at the tips of the lobes and potential lobes. The furrows (numbered in sequence of origin) are regions where wall material does not accumulate. The outlines represent successive stages during growth of the semicell. i = isthmus. (Kiermayer 1981)

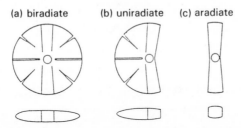

Figure 6.17 Effects of irradiation on *Micrasterias* development. When only the cytoplasm is irradiated, growth is aberrant (a) but not permanently impaired because after several divisions the daughter semicells become normal shape. If the nucleus is irradiated there can be permanent loss of lobes so that uniradiate (b) and aradiate (c) cells are formed. (Kallio & Lehtonen 1981)

lobes of the semicell and ultimately by the nucleus. If a growing lobe is irradiated with a UV microbeam, the growth of that lobe may be stunted, even for several generations, but eventually the daughter cells can resume the normal shape, showing that the information for shape production has not been lost. On the other hand, irradiation or other treatment of the nucleus can sometimes result in the permanent loss of a

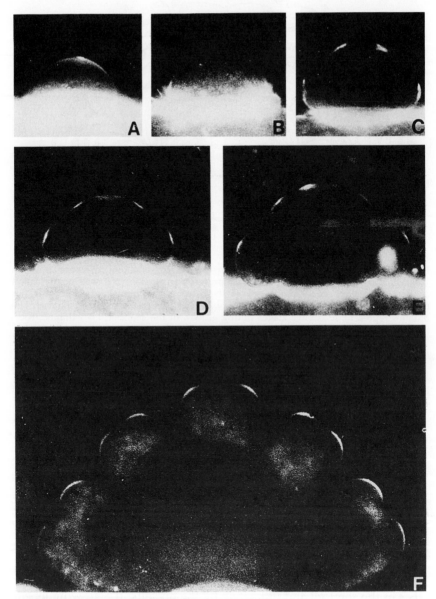

Figure 6.18 Distribution of bound Ca^{2+} in growing *Micrasterias* semicells. Cells treated with chlorotetracycline fluoresce where it binds to Ca^{2+}, which is localized to the tips of the lobes where wall accumulation is greatest. (Meindl 1982; photographs kindly supplied by Dr U. Meindl)

lateral lobe to produce a uniradiate form (Fig. 6.17), which is then perpetuated in the same way as though it is a mutation.

The cytoplasmic basis for the patterning has been looked for but so far nothing structural has been found, although membrane-bound Ca^{2+} is distributed in a way that corresponds to the expected pattern (Fig. 6.18). Plasmolysis shows that the plasma membrane adheres strongly to the wall at the sites of the furrows and invaginations, but not at the sites where the cell wall material accumulates in turgor-stressed cells.

During growth and patterning of the growing semicell, the nucleus continually provides necessary information. If the nucleus is destroyed, the semicell develops as in anucleate cells, which can also be produced by centrifuging the nucleus into the mother cell just before division so that both daughter nuclei become confined to the mother cell and the new semicell is anucleate. The anucleate semicell then forms only three simple lobes, corresponding to the polar lobes and the two lateral lobes, but there is no further elaboration of the cell pattern. No secondary wall forms and the primary wall soon expands and bursts.

If the nucleus is irradiated at various times during the growth of a new semicell, the effects of nuclear disturbance are seen in a disturbance of semicell growth 30–155 min later. The earlier the nucleus is irradiated the greater the subsequent disruption of semicell growth. This means that the nucleus is apparently continuously supplying messages that control semicell growth. The same sort of disruption or inhibition of semicell growth can be produced by treatment of the cells before division and septum formation with actinomycin D (which inhibits RNA synthesis), with puromycin (which inhibits the translation of RNA during protein synthesis), or by irradiation of the cytoplasm of the developing lobe with UV light. All of these treatments mimic enucleation and imply that the effect of the nucleus on semicell growth is mediated by the RNAs that it produces (see also *Acetabularia*, Ch. 12, Box 12.1).

In the growth of the new semicell, the cell wall material is apparently carried to the sites of wall synthesis by vesicles formed from the dictyosomes, as in other plant cells. When wall synthesis first begins, the vesicles are large and contain pectic substances or are concerned with plasma membrane synthesis. Then, small vesicles are formed, which are concerned with the growth of the primary wall and cellulose synthesis. Later, the formation of the secondary wall, which consists of layers of lamellae at 120° to each other, is associated with the formation of flat vesicles by the dictyosomes. When the nucleus is irradiated, the formation of the next set of vesicles is prevented. There is also a reduction in the formation of vesicles of any sort by the dictyosomes. The sequence of different sorts of vesicles produced by the dictyosomes is, therefore, under nuclear control. The control of wall synthesis by the nucleus is, therefore, exerted indirectly through the dictyosomes.

127

The movement of dictyosome vesicles is not greatly affected, nor is cell shape affected, by disruption of microtubules (e.g. by colchicine). Disruption of microfilaments (by cytochalasin B) does, however, stop cytoplasmic streaming and expansion of the primary wall. Although microtubules are not implicated in determining cell shape, they are involved in anchoring the nucleus at the isthmus (cf. fern protonemata, see section 6.5.1) and in the movement of the nucleus and the chloroplast during cell division.

In *Micrasterias*, cell shape therefore depends on differential wall synthesis according to a predetermined pattern somehow present in the plasma membrane. Nuclear control is exerted through the RNAs that it produces and they act on cell shape 0.5–2 h after they have been produced by the nucleus. Nuclear control is also exerted by the control of the type of vesicles produced by the dictyosomes and the sequence in which they are produced. Microtubules, in *Micrasterias*, are not involved in the determination of cell shape.

6.7 SUMMARY

(1) Ion currents are an apparently ubiquitous accompaniment of polar growth. An influx of calcium precedes and accompanies localized growth of plant cells and calcium tends to accumulate at the point of entry, i.e. the absorptive pole. Calcium efflux occurs either basally in the same organ or at the opposite, photosynthetic pole. A gradient of calcium concentration becomes established between influx and efflux sites. In aerial organs, the current may be carried beneath the cuticle instead of in an external medium. Ion currents as small as 20nA cm^{-2} can be measured with the vibrating probe electrode.

(2) The ion currents are caused by a redistribution or localized activation of ion channels and pumps in the plasma membrane. The currents can often be induced by unidirectional external stimuli. After ion currents have been set up, the cytoplasmic organelles become redistributed, perhaps in part by electrophoresis within the cell, and the wall begins to grow locally at the site of current entry.

(3) The establishment of polarity has been studied particularly in zygotes of the seaweeds *Fucus* and *Pelvetia*. The polar axis is labile for the first 12 h, during which it can be reorientated by unidirectional light. The axis then becomes fixed. Inhibitors of axis fixation lengthen the period over which the axis can be made to reorientate. An effective inhibitor of axis fixation is cytochalasin B, indicating the involvement of actin microfilaments of the cytoskeleton in axis fixation, perhaps to establish a directional subcellular transport system.

(4) Outgrowth of the *Fucus* rhizoid is preceded by movement of dictyosome vesicles to the rhizoid pole and the incorporation there of a sulphated fucoidin into the wall. Sulphation of this carbohydrate seems necessary for its localized accumulation, but this seems to be a consequence of polarity rather than a cause.

(5) When the polarity of growth changes, as in *Vaucheria* branching or the transition to two-dimensional growth in fern protonemata, there are first changes in the distribution of current flowing through the cells so that a new point of current entry is established at the new growing tip. This is followed by changes in wall structure, which allow or cause a new direction of cell growth.

(6) Unequal cell divisions precede and seem necessary for cell differentiation. They are the result of either the symmetrical division of a cell in which the cytoplasm is already polarized, or the asymmetrical division of a non-polarized cell. In both cases, this results in daughter cells with different cytoplasmic constitutions.

(7) Unequal divisions are often preceded by movement of the nuclei in the dividing, and also in neighbouring, cells. Before unequal division in the germinating fern spore, the nucleus migrates to one side of the cell and then to one end, where it divides. The side of the cell to which it first migrates is next to the site of heavy metal binding. The ability to bind ions at this site changes during the cell's development. It is probably here that ion pumps and channels are localized in the cell membrane. The nucleus moves towards the binding site during the second peak of binding ability. The natural ion thought to be bound is Ca^{2+}.

(8) In tip-growing cells, the nucleus migrates forward at the same rate as the tip and so remains a fixed distance behind it. The nucleus is probably moved and held in position by microtubules, which hold the nucleus more firmly just before division and at the division site. Nuclei displaced basally by centrifugation migrate back to their 'proper' position unless displaced too far just before division, in which case they divide in a basal location in the filament and a side branch forms.

(9) In tip-growing cells growing rapidly, cell wall extension occurs just at the rounded tip. The cytoplasmic organelles have a characteristic distribution: at the tip there are many dictyosome vesicles, further back there are mitochondria and (except in fungi) plastids. More basal still is the vacuole, and still further back is the nucleus. The cytoskeletal basis for this organization is as yet unknown.

(10) In the unicellular desmid, *Micrasterias*, each cell consists of two distinctively shaped semicells connected by an isthmus, in which lies the nucleus. After division, each semicell forms a new semicell, normally with a pattern that is the mirror image of that of the

129

parent semicell. The growing semicell becomes lobed by the greater rate of wall growth at the tips of the lobes and less or no growth in the furrows. This can be made visible as localized wall thickenings ('stored growth') when cells are placed in an osmoticum that temporarily prevents cell expansion. The nucleus is necessary for wall growth. Irradiation of the nucleus affects the growth of the wall about 30 min later. The influence of the nucleus is partly mediated through RNA and partly through its control of the type of dictyosome vesicle formed at successive stages of cell development. The basis for the cytoplasmic pattern, which specifies the position of the lobes and furrows and which seems to be imprinted on the plasma membrane, has not yet been discovered. In *Micrasterias*, microtubules do not seem to be involved in the determination of cell shape and polarity.

FURTHER READING

Bentrup, F.-W. 1984. Cellular polarity. In *Encyclopedia of Plant Physiology*, New Series, Vol. 17, H. F. Linskens & J. Heslop-Harrison (eds), 473–90. Berlin: Springer.

Cooke, T. J. & R. H. Racusen 1986. The role of electrical phenomena in tip growth, with special reference to the developmental plasticity of filamentous fern gametophytes. *Symposia of the Society for Experimental Biology* **40**, 307–28. (Ion currents in fern protonemata)

Kiermayer, O. (ed.) 1981. *Cytomorphogenesis in plants*. Vienna: Springer.

Quatrano, R. S. 1978. Development of cell polarity. *Annual Review of Plant Physiology* **29**, 487–510.

Quatrano, R. S., S. H. Brawley & W. E. Hogsett 1979. The control of polar deposition of a sulfated polysaccharide in *Fucus* zygotes. In *Symposia of the Society for Developmental Biology*, Vol. 37, S. Subtelny & R. Konigsberg (eds), 77–96. New York: Academic Press.

Schnepf, E. 1986. Cellular polarity. *Annual Review of Plant Physiology* **37**, 23–47.

Weisenseel, M. H. & R. M. Kicherer 1981. Ionic currents as control mechanisms in cytomorphogenesis. In *Cytomorphogenesis in plants*. O. Kiermayer (ed.), 379–99. Vienna: Springer.

NOTES

1   Jaffe (1966), Jaffe (1969), Robinson & Jaffe (1975). (Polarity in algal zygotes)
2   Jaffe & Nuccitelli (1974). (Vibrating probe electrode)
3   Quatrano *et al.* (1985), Kropf *et al.* (1988, 1989). (Axis fixation in algal zygotes)
4   Dyer & King (1979), Miller (1980), Stetler & Demaggio (1972). (Fern gametophytes)
5   Bassel (1985), Kotenko *et al.* (1987), Miller (1985), Robinson *et al.* (1984). (*Onoclea* fern spore germination)

6 Mineyuki & Furuya (1986), Mineyuki & Furuya (1980), Reiss & Herth (1978; 1979), Sievers & Schnepf (1981), Wada *et al*. (1980). (Tip growth)
7 Kallio & Lehtonen (1981), Kiermayer (1981), Meindl (1982), Selman (1966), (*Micrasterias*)

PART IV

Cell differentiation

CHAPTER SEVEN

Control of the differentiation of vascular tissues

7.1 REGENERATION OF VASCULAR TISSUES IN WOUNDED PLANTS[1]

In animals, there are probably about 100 or more different cell types. In plants, there are fewer, probably about 40 (Table 7.1). Plant cell types are distinguished by cell wall structure, shape, size and position of the cells, and cell contents. The reactions of the cells to the common histological stains are helpful here. Lignified cell walls, for example, stain intense red with safranin. Because of their frequently larger size and characteristically thick, pitted walls, xylem elements are often the most easily recognized. Phloem cells are less easily recognized, but can be detected by stains specific for callose, which is a carbohydrate that forms on phloem sieve plates. Because they are organized into bundles or strands, vascular tissues – especially xylem and phloem – are often the most obvious tissues. Also, as their development has proved to be amenable to experimental manipulation, their differentiation has been studied much more than that of other cell types.

When a plant stem is wounded by a cut that severs a vascular bundle, new vascular tissue is formed that bypasses the wound and links up the severed tissues above and below the wound. This system has been exploited to study the regeneration of vascular tissues, especially in *Coleus* where the stem is square in section, the leaves are inserted on the flat faces of the stem, and the arrangement of the vascular tissues in relation to the leaf positioning and the stem geometry is convenient for experimentation. At each corner of the stem there is a large vascular bundle, so a cut on the corner of the stem severs a main vascular bundle (Fig. 7.1). There are smaller vascular bundles in the centre of each flat face of the stem, and between these bundles and the corner bundles are smaller bundles that consist only of phloem.

The regeneration of new vascular strands round the wound has

135

Table 7.1   Plant cell types.

meristematic apical cells	root cap cells
parenchyma	root hairs
collenchyma	cambium
epidermis	ray cells
endodermis	cork
pericycle	phelloderm
fibres (sclerenchyma)	cork cambium
stomatal guard cells	laticifers
stomatal subsidiary cells	secretory cells
palisade	hairs
mesophyll	egg cell
xylem vessels	synergids
tracheids	antipodal cells
phloem sieve tubes	endosperm cells
companion cells	aleurone
transfer cells	tapetum
gland cells	pollen generative cell
idioblasts	pollen vegetative cell
cystoliths	

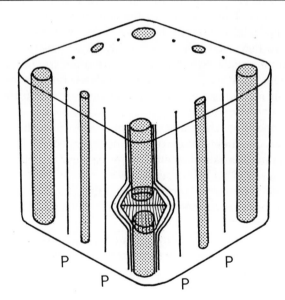

Figure 7.1   Regeneration of severed vascular bundles. A small wedge cut out of the corner of a *Coleus* stem severs one of the four main vascular strands. The smaller strands, some of which (P) contain only phloem and no xylem, can be severed by appropriately positioned cuts on the face of the stem. Linking strands regenerate round the wound.

usually been followed by observing the formation of xylem tracheary elements, which are the easiest cells to see and identify in cleared stems. Jacobs showed that the formation of new vascular strands in *Coleus* has the following characteristics:

(1) it progresses mainly basipetally, though some regeneration is acropetal (from the base of the wound upwards);
(2) it depends on the presence of an adjacent leaf above the wound (except for acropetal differentiation);
(3) it can also occur when an excised leaf is substituted by applied auxin, and the effect of the leaf can be duplicated by auxin alone.

The auxin produced from the basal cut ends of excised leaves was collected in agar blocks and the amount was measured by bioassay. The amount of auxin equivalent to that produced by a single leaf was then supplied to the **petiole** stump where the leaf had been cut off. This amount of applied auxin mimicked the leaf in causing the same amount of xylem to form round the wound made in the stem below the excised leaf (Table 7.2).

These experiments show that auxin applied at relatively high, and almost certainly unphysiological, concentrations can substitute for the leaf in the regeneration of xylem in *Coleus*. The simplest hypothesis is that the leaf is a source of auxin and that this is the sole limiting factor, and hence the controlling factor, in xylem formation in the *Coleus* stem. When phloem regeneration was also examined, it was found that this too could be controlled by auxin application. These experiments are consistent with auxin produced by the leaves being all that is necessary to cause xylem and phloem regeneration around wounds that sever vascular bundles. In those bundles consisting only of phloem and having no xylem, it was shown that phloem regenerating alone was also dependent on auxin.

When this system for studying vascular regeneration was simplified, by isolating the wounded internodes and culturing them with their ends in agar, there was an additional requirement for cytokinin and sometimes for sugar. These requirements, and sometimes others as well, presumably reflect the fact that tissues normally have these substances supplied to them from elsewhere in the intact plant. In isolated internodes or, as we shall see later, in callus or cell cultures, differentiation of vascular tissues may, therefore, be limited and controlled by various substances that, in the intact plant, are apparently in adequate supply. But in *Coleus* plants that are intact except for the wound in the stem and the removal of one or a few leaves, and kept in the light, auxin appears to be able to act as the main controlling factor for differentiation of xylem and phloem. Auxin can act as a controlling factor in these experiments; whether it actually does so in the intact plant is still open to question.

Table 7.2 Auxin can substitute for leaves and buds in the regeneration of vascular strands around a wound in a *Coleus* stem. There were significantly more strands formed when leaves were present above the wound or replaced by auxin than when these upper (distal) leaves were removed. There was no statistically significant reduction in the numbers of strands formed when auxin substituted for the leaves and buds. Eleven plants per treatment. (Jacobs 1979)

	Average number of xylem strands formed ($\pm$ SE)	
control plants (all leaves and buds left on)	15.0 ± 1.3	
		Not statistically significantly different
leaves and buds above the wound excised and replaced by 2 mg l^{-1} IAA	11.5 ± 1.2	
		Significantly different at $p < 0.01$
leaves and buds above the wound excised	6.3 ± 1.4	

7.2 FORMATION OF XYLARY ELEMENTS IN CULTURE AND CELL SUSPENSIONS[2]

In order to study the control of differentiation more readily, tissue and cell cultures have been used to avoid the complications of the differentiating cells being part of a larger cell mass in which some cells are already differentiated. In tissue or cell cultures of apparently homogeneous and parenchymatous cells, the processes of differentiation are studied more easily. Another advantage of such systems is that the progress of differentiation can be timed, from the start of the culture. The cells that have been studied most often are xylem elements because they are recognized most easily.

Cultured cells and tissues have been especially useful in tackling the questions:

(1) Is it necessary for the cells to synthesize DNA or divide before they can differentiate, or can cells directly transform from one cell type into another without cell division?

(2) What stimuli are required for differentiation?

138

7.2.1 Is cell division or DNA synthesis necessary for cell differentiation?

Usually, when explanted tissues are cultured there is some cell division, even if it is just the result of the wounding incurred in cutting out the explant, so cell division normally precedes any subsequent differentiation. In the intact plant, differentiated cells have necessarily been derived by division from common, undifferentiated ancestral cells. In order to test whether cell division itself is essential for differentiation, apart from being the mechanism for generating cells, it is necessary to inhibit division completely in the explants or cell suspensions and to see whether cell differentiation can still occur.

In many pith tissues there are isolated xylem elements that are the same shape and size as the surrounding cells. Their appearance suggests that they have not been formed by a recent cell division but are pith cells that have differentiated to form tracheary elements, recognized by their highly thickened and sculptured walls (Fig. 7.2). However, especially when they form strands, it is not always clear whether division has indeed taken place to produce the xylem cell and another, less obvious, cell alongside or behind it. Formation of xylem elements without cell division has been claimed for pea (*Pisum*) pith explants, but the data are not rigorous enough to rule out cell divisions.

Explants of Jerusalem artichoke (*Helianthus tuberosus*) tuber pith consist of homogeneous parenchymatous cells all in the 2C (G_1) phase of the cell cycle. Growth and cell differentiation in these explants normally requires cell division, which can be promoted by the addition of auxin to the culture medium. But if the explants are first γ-irradiated, cell division and DNA synthesis can be prevented completely but xylem cell differentiation can still occur, although reduced in amount and later than in the unirradiated controls.

A system that has been exploited in recent years is *Zinnia* mesophyll cell suspensions. If *Zinnia* leaves are ground up in a pestle and mortar in culture medium, a suspension of intact cells is obtained and the mesophyll cells can be separated from the rest by filtration. If the mesophyll cells are cultured in the presence of auxin and cytokinin, then 60% of the cells transform into xylary elements (Fig. 7.3). This still happens even if cell division is completely inhibited with colchicine or γ-irradiation. Cell division is, therefore, not essential for differentiation in these cells. However, it has been argued that differentiation can only proceed if the cells have undergone some critical biochemical events during the first part of G_1 in the cell cycle. If so, then mesophyll cells that happen to be in the first half of G_1 would be able to differentiate immediately. Those in the second half of G_1 would have to go round the cell cycle and divide in order to get to the critical, early part of G_1 before they could differentiate.

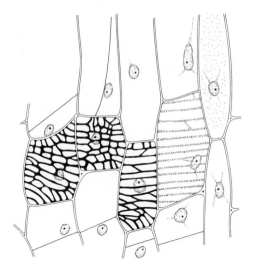

Figure 7.2 Formation of tracheary cells in pith. Cells in the outer layers of the pith (between vascular bundles) that have differentiated into xylem tracheary elements. The cells appear to have transformed directly without division, but divisions parallel to the plane of the section may have occurred but not been detected, and divisions are seen in adjacent cells. (Sinnott & Bloch 1945)

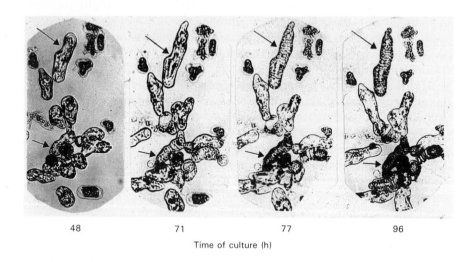

| 48 | 71 | 77 | 96 |

Time of culture (h)

Figure 7.3 Differentiation of *Zinnia* mesophyll cells into xylem cells. In culture for 48 h, 71 h, 77 h, and 96 h, the arrowed cells transform into xylem elements with characteristically thickened walls. (Fukuda & Komamine 1980; photograph kindly supplied by Professor A. Komamine)

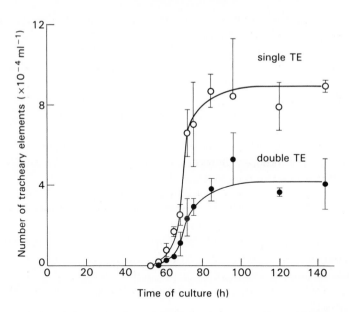

Figure 7.4 Differentiation of *Zinnia* mesophyll cells with different starting amounts of DNA. Single tracheary elements (TE) had the 4C amount; double TE had divided to form a cell pair and each cell had the 2C amount of DNA. 2C and 4C cells all developed simultaneously. (Fukuda & Komamine 1981)

This would then explain why 40% of the *Zinnia* cells did not differentiate and would also explain why cells sometimes seemed to require cell division before differentiation and sometimes not.

Further experiments with *Zinnia* cells have shown that this is not a sustainable hypothesis. Cells that had first divided differentiated at exactly the same time as those that had not divided, and in fact differentiation became visible at about 6 h after division (Fig. 7.4). If these divided cells had had to traverse an extra bit of G_1, it would have been expected that they would have been delayed in differentiating, but they were not. The simplest hypothesis is that differentiation and progression through the cell cycle are two parallel but independent processes. However, inhibitors of DNA synthesis, including fluorodeoxyuridine, fluorouracil, and aphidicolin, inhibited differentiation. Some inhibitors would have other effects, such as inhibiting carbohydrate metabolism and interfering with wall thickening, but aphidicolin seems to specifically inhibit DNA polymerase. Synthesis of some minor DNA component, or DNA repair processes, may, therefore, be essential for differentiation.

Whether this is generally true for other cells and systems has yet to be shown. Since determination and, therefore, differentiation can be regarded as a stepwise process (see Ch. 10), it could be argued that a young mesophyll cell requires very little adjustment to become xylem.

After all, it is in the potential mesophyll cells of the developing leaf that the veins of the leaf differentiate, apparently under the influence of auxin (see section 7.6.2). Whether less-differentiated cells, such as callus, which has been parenchymatous for many cell generations, or the cells of the young globular embryo can become xylem elements (or any other highly differentiated cell type) by direct transformation without intervening, and possibly essential, cell divisions is much more open to question.

Because some cell types can differentiate into xylem elements without cell division immediately beforehand, this does not necessarily imply that cell division is not required at earlier stages of differentiation in order to create neighbouring cells with different developmental potentialities. It seems unlikely, for instance, that differentiation without cell division could give rise to phloem cells or to stomatal guard cells. We must bear in mind that the *Zinnia* mesophyll cell suspension may not be typical. It has been used precisely because it can be manipulated so that cell division and cell differentiation can be separated. There may still be many systems in which cell division is essential, although it may be difficult to prove this.

7.2.2   What stimuli are required for differentiation?

The stimuli required for differentiation are found by experiments in which various components of the inducing medium are omitted or their action prevented by the use of specific inhibitors also incorporated into the medium. In *Zinnia* cell suspensions, differentiation does not take place unless the media contain both auxin and cytokinin in addition to basal salts, cofactors, and sucrose. Omission of any of these can reduce or prevent differentiation. However, the requirement for these substances for differentiation has not been tested directly in the absence of cell division, e.g. in irradiated cells. Irradiated cells would seem to provide a useful system to try to find which components of the medium are required specifically for differentiation where this process can be isolated experimentally. In the whole plant, where the plant itself presumably supplies all the accessory but essential substances, only auxin seems to be required specifically. The answer seems to be that many substances, required for growth, are required for differentiation but that various substances can inhibit the process if they are in short supply, and auxin can perhaps control it.

7.3   INDUCTION OF VASCULAR TISSUE IN CALLUS: FORMATION OF NODULES[3]

Wetmore and colleagues showed that when a bud was grafted into a block of lilac (*Syringa*) callus, scattered nodules of vascular tissue formed in the callus below the bud. The inference was that the bud was producing some substance that induced the differentiation of vascular tissues in the callus (Fig. 7.5). When a mixture of auxin and sucrose alone was substituted for a grafted bud, nodules of vascular tissues differentiated in the callus. Moreover, the nodules formed a ring some distance below the point of application of the auxin and sugar. Other vascular nodules also formed at the base of the callus near the medium on which the callus was cultured. When the auxin and sucrose were supplied continuously from a pipette, a ring of vascular tissue formed rather than discrete nodules. With increasing auxin concentration, the ring of vascular tissue was wider and further from the source. At high concentrations of auxin and sucrose, the ring of vascular tissue was not formed at all, as though it could only be formed outside the callus block, which was not wide enough to contain it.

Altering the concentration of sucrose had no effect on the position of the differentiating vascular tissues, but with 1–2% sucrose, xylem predominated and little or no phloem was detected. With 2.5–3.5% sucrose, both xylem and phloem were present in the nodules, with the xylem innermost and cambium between the xylem and phloem, as in the normally expected orientation. With 4% sucrose, only phloem and little or no xylem was formed.

From these experiments, several important conclusions can be drawn:

(1)   Differentiation of vascular tissues in callus can be induced with simple chemicals: auxin and sucrose.

(2)   The vascular nodules form at a specific distance from the auxin source, depending on the concentration of the supplied auxin.

(3)   Increasing sucrose concentrations altered only the proportions of xylem and phloem and so could control the composition of the vascular tissue.

(4)   Auxin and sucrose can substitute for a bud, suggesting that a bud could control vascular differentiation simply according to how much of these substances it produces.

(5)   The orientation of the induced vascular tissues is the same as in the intact plant.

These experiments are, therefore, consistent with (but do not prove) the fact that the bud is a source of auxin and sucrose, which control the induction of differentiation of xylem, cambium, and phloem. However, several questions are raised:

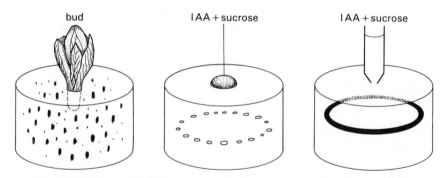

Figure 7.5 Induced differentiation of vascular tissue in callus. (a) A bud grafted into a block of *Syringa* callus induced the formation of vascular nodules. (b) A mixture of only auxin (IAA) plus sucrose induced a ring of vascular nodules. (c) A continuous supply of auxin from a pipette induced a continuous ring of vascular tissue. (After Wetmore & Rier 1963)

(1) Buds may be sources of auxin but are they likely to be sources of sucrose?

(2) Why are nodules formed rather than a continuous ring of tissue, i.e., why do the cells between the nodules not differentiate into vascular tissues?

(3) Why are nodules formed only at a particular distance from the auxin source, i.e., presumably only at a specific point on the diffusion gradient and at a specific concentration of auxin?

(4) Why are not vascular strands formed, as in the intact plant?

The answer to the second question is given partly by the experiment with the pipette, showing that where a continuous supply of auxin, and presumably a higher concentration, is given, then a ring of tissue forms rather than nodules. These may form only when the supply is insufficient to prevent it from becoming exhausted before all cells can be triggered. However, this does not answer the point that the arrangement of the nodules is relatively regular; this is discussed further in Chapter 12.

The formation of vascular tissues only at a particular distance from the auxin source implies that the cells are induced to differentiate only at specific concentrations of the inducer. This raises questions about the sensitivity of the cells and their content of receptors. How the filling of a receptor site leads to the differentiation process is one of the great mysteries of cell biology. A related point is why cells differentiate as either xylem or phloem. Is the difference in concentration of inducer between two adjacent sites sufficient for it to be sensed and for the cells to react accordingly? Or is vascular tissue specified and the different cell types within it always developed together or in a specific order? This question will be considered a little later (see section 7.5) with experiments showing that xylem does not normally seem to form without phloem, although phloem can form by itself. The problem may be that, in reality, it is procambium differentiation that is induced in callus, the differentiation of the various vascular cell types from it being modified, but not induced, by substances such as sugars and growth substances. However, the prior differentiation of procambium may not always occur, as it evidently does not in the isolated *Zinnia* cells.

The fourth question – Why are strands of vascular tissue not formed? – highlights the difference between callus and the intact shoot apex. In the shoot, cells are continually being added apically by the growth of the apical meristem. Essentially, nodules of vascular tissue may be induced, as in the callus, but with the locus of induction remaining at a set distance below the apical meristem, which continually moves upward as it produces cells at its base. The implication here is that it is the apical meristem that is the source of inducer. For auxin, this may be plausible, but it is less so for sugar and would be consistent with the concentration of auxin being the important factor. Other substances would act as inducers only by virtue of being enabling factors, and would not give the positional effects in the way that auxin may do.

7.4   DOES CELL DIFFERENTIATION REQUIRE SPECIFIC CONCENTRATIONS OF INDUCERS?[4]

If the concentrations of the inducing substances are important, can it be shown:

(1)   what concentrations are required for induction of different types of vascular tissues;
(2)   whether differentiation of one cell type can occur in the absence of other cell types;
(3)   if more than one cell type is differentiated, is there any particular sequence in which this occurs?

By applying radioactive auxin (IAA) and sucrose to the surface of blocks of bean (*Phaseolus*) callus, it was possible to measure their concentrations at the site of vascular differentiation (Fig. 7.6). The values were 25 μg l^{-1} for IAA and 7.5 g l^{-1} for sucrose. The implication is that these are concentrations that in some way trigger the differentiation of vascular tissues in bean callus. However, this experiment would bear repeating to show that there is indeed a gradient of auxin concentration through the callus and that the concentration of auxin at the site of differentiation really mattered.

We do not know whether in different tissues or species the concentra-

145

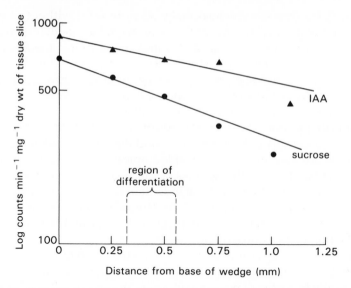

Figure 7.6 Concentrations of auxin (IAA) and sugar at the site of vascular differentiation in *Phaseolus* callus. Differentiation of vascular nodules was induced by auxin + sucrose supplied in an agar wedge. In parallel experiments, each was radioactively labelled. After 2 days, the callus was chopped up and assayed. The concentration of auxin and sucrose at the site of vascular differentiation could then be measured. (Jeffs & Northcote 1967)

tions would be different. We would expect so, since the requirement for these substances presumably indicates that in this system they are in limiting amounts. In other systems, they may be either more, or less, limiting, depending on the endogenous concentrations in the experimental tissues. The concentrations of IAA and sucrose measured in the bean callus may, therefore, mean very little in absolute terms. Even if the differentiation of vascular tissue did occur only at a specific auxin concentration, sucrose would necessarily also be at some definable, but not necessarily critical, concentration at that same point in the callus.

The concentration of IAA and sucrose could be interpreted as decreasing exponentially from the source. If so, this would be consistent with the movement of IAA and sucrose through the bean callus being non-polar and by simple diffusion. This would explain why TIBA (triiodobenzoic acid, an inhibitor of polar auxin transport) was without effect on differentiation at concentrations close to or less than that of the IAA. (In *Zinnia* cell suspensions, TIBA has to be supplied at five times the auxin concentration in order to delay and reduce the amount of differentiation, and at fifty times the auxin concentration to inhibit differentiation.)

The huge difference in concentration ($>10^5$ times) between auxin and sucrose at the differentiating site in the bean callus may suggest that the sucrose is simply providing a source of carbon for the tissue. However, other sugars could not be substituted – only sucrose (and, to a lesser extent, maltose and trehalose) could induce vascular tissues. Sucrose, as well as IAA, may, therefore, be acting as a specific inducer. The sucrose was effective in inducing complete nodules of vascular tissue only if it was applied simultaneously with, or following, the IAA. Sucrose followed by IAA gave only some xylem **tracheids**, the same as with IAA alone. Sucrose may, therefore, be modifying the development of cells produced in the presence of IAA, which appeared to give some stimulation of cell division. The addition of cytokinin to the sucrose and auxin stimulated further cell division and resulted in an increase in phloem differentiation.

In *Todea* (a fern), the prothallus does not normally make xylem. Addition to the growth medium of any one of sucrose, auxin, cytokinin, or gibberellin caused differentiation of xylem tracheids; therefore, no single substance *per se* was a critical limiting factor. In *Parthenocissus* callus, with auxin at 0.1 g l^{-1} in the medium, there was no vascular differentiation. It required sucrose and, solely by increasing the sucrose concentration, xylary elements were induced. The higher the sucrose concentration (up to 8%), the greater the amount of xylem formed and the larger the arcs of xylem in the callus. These xylem arcs were unusual in that they had a cambium *internal* to them. Phloem was not observed but could have been present, although it is not obvious where. In this callus, auxin was required but differentiation was experimentally induced by varying the sucrose concentration.

While it may be argued that there is always sufficient auxin present endogenously in those tissues in which sucrose appears to be the controlling factor, and *vice versa*, it is difficult to argue that there is any *single* controlling factor; different substances may be so in the appropriate circumstances.

7.5   DIFFERENTIATION OF XYLEM AND PHLOEM: CAN EITHER FORM ALONE?[5]

The differentiation of one cell type, xylem, can certainly occur in the absence of phloem in *Zinnia* cell suspensions. In intact plants, vascular strands may sometimes consist solely of phloem, as in the xylem-less strands in *Coleus*, but strands solely of xylem do not form in the absence of phloem. Often, only xylem elements are recorded in experiments on vascular differentiation because these are the most easily recognized; in the intact plant, however, phloem always differentiates before xylem is formed. Phloem differentiation always seems to require cell division, perhaps because the unit in the phloem is the sieve tube plus companion cell. During the differentiation of vascular tissues in wounded *Coleus*

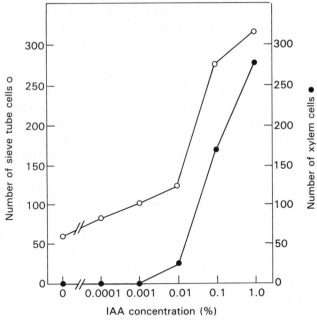

Figure 7.7 Differentiation of xylem and phloem around a wound in *Coleus* stems. In vascular strand regeneration, xylem requires a higher auxin concentration than phloem. (Jacobs 1979)

internodes, separation of the xylem from the phloem by a coverslip barrier resulted in xylem regeneration only on the phloem side of the coverslip, suggesting that phloem had to be present first in order for new xylem to be formed. This is consistent with the dependence of both xylem and phloem formation on auxin concentration, but with phloem forming at lower concentrations than xylem (Fig. 7.7). However, strands of **fibres** may differentiate in the absence of both xylem and phloem. Whether the differentiation of xylem tracheary elements in the pith of intact plants also occurs in the absence of phloem is less clear, although the apparent direct transformation of pith cells into tracheary elements has been observed in a number of cases.

In tissue cultures of several species, Aloni showed that xylem tracheary elements never differentiated in the absence of phloem. Xylem and phloem are probably always formed together in callus. Even in the *Parthenocissus* callus, phloem may have been there but not observed because it was not specifically looked for, and it may perhaps have been in an unusual orientation, internal to the xylem. But in tissue cultures and cell suspensions, we have seen that differentiation of xylem elements can certainly occur by direct transformation of the cells in the absence of phloem, as in *Zinnia*. There are apparently two types of system in which vascular cells differentiate. The first is the multicellular system of the intact plant or callus in which vascular strands or nodules are formed and in which phloem is always formed first (and sometimes alone, as in xylem-less strands). The second is the disorganized system of isolated cells in which the cells can transform directly into tracheary elements. The probability is that cell division always takes place in the multicellular system, and is involved with phloem formation, but in the isolated cells, cell division is not essential. In isolated cell suspensions, there is no evidence of phloem being formed at low concentrations of auxin. Indeed, even if it did occur it would have to be in different cells from those forming xylem. A possible explanation for the need to supply cell suspensions, but not multicellular systems, with cytokinin is that in the multicellular systems it is normally supplied in sufficient quantity by the phloem as it differentiates. This remains speculation in the absence of direct measurements.

7.6   INDUCTION OF VASCULAR STRANDS[6]

In the cell cultures and suspensions, there is differentiation of individual vascular cells. In callus, there is differentiation of vascular nodules or rings. In lettuce pith explants, short vascular strands form just below the surface of the explant but are not continuous from one end of the explant to the other. Only in the whole plant are vascular strands formed, which link up with each other to form a vascular network. Vascular strands differentiate into or from young developing leaves, which are believed to be sources of auxin. It was, therefore, natural to ask whether the apparent directing effect of young leaves on the differentiation of vascular strands could be replaced by auxin.

First, we should consider the effect of simply removing the young leaves as soon as each begins to form. When this is done, the vascular traces normally formed below the leaves do not differentiate. In ferns in which the leaf primordia were removed as they were initiated, the formation of the leaf trace vascular network was suppressed and only the apex-induced **solenostele** vascular tissue was formed. In lupin (*Lupinus*), the leaves determined their traces, and although auxin could not substitute for leaves in causing complete differentiation of the vascular bundles, it was able to prevent the procambium from becoming parenchymatous as it did when the leaf was removed. In *Coleus*, replacement of a very young leaf primordium with an auxin bead replaced the effect of the leaf in inducing vascular strands below it in the stem.

In the pea (*Pisum*), not only are complete vascular strands induced below each leaf but also a strand consisting solely of fibres. When the leaf is removed this fibre strand does not form, and when the young leaf is split, two strands form. Only young leaf primordia less than 75 μm long are effective in promoting fibre differentiation, whereas older leaf primordia promote vascular strand differentiation. Applied auxin (IAA) could substitute for a leaf by inducing vascular strand differentiation but not fibre differentiation. Fibre strands joined only to other fibre strands, and vascular strands only to other vascular strands. These various observations were taken to mean that the stimulus for the differentiation of vascular strands and fibres was different, and that fibre differentiation requires something either in addition to auxin or altogether different. However, the principal auxin in the pea is phenylacetic acid and its effect on vascular strand differentiation in peas has not been tested. When several natural auxins are found in the same plant, we need to know more about their relative activities before drawing the negative, and possible premature, conclusion that auxin is not involved.

The induction of new vascular strands in relation to the existing strands has been studied by T. Sachs. The vascular strands from a leaf enter the stem at the node at which the leaf is inserted but join with the vascular tissue in the stem only at a lower node. Thus, there is the characteristic leaf gap, where the stem vascular bundles become parted to allow 'entry' of the strand from the leaf (Fig. 7.8a). What actually happens during development is that when a new leaf is formed at the shoot apex, its procambial strand differentiates and leads down through the incipient petiole into the stem. At the same time, vascular strands from younger leaves, just above, also begin to differentiate. As the younger, more distal, strands differentiate they avoid and do not join immediately with the slightly older strands from the leaves just below them. They join only lower down the stem than the node at which the older leaf is inserted.

If the leaf of a pea is excised while its vascular trace is still forming, this leaf trace is then joined by the stem vascular tissues at the node of that leaf (Fig. 7.8b) and not at the node below (Fig. 7.8a). This shows that removal of the leaf, the source of induction for its strand, also removes the inhibition preventing the apical strand from fusing with it. Similarly, when the apical bud is excised and an axillary bud grows out (because of the removal of apical dominance), its differentiating vascular strand will join with the stem vascular bundles (Fig. 7.8c). If both leaf and apical bud are excised, the axillary bud strand may join with both leaf and stem strands (Fig. 7.8d). If the apical bud is not excised but is damaged sufficiently to reduce its apical dominance, then this allows the outgrowth of the leaf's axillary bud. The apical bud still remains as a source of induction for vascular tissues, so that when the leaf is excised the vascular strand induced by the growing axillary bud joins with the strand from the excised leaf but not with the strand from the apical bud (Fig. 7.8e). These results are consistent with the leaf, the growing axillary bud, and the apical bud all being sources of induction for vascular strands.

150

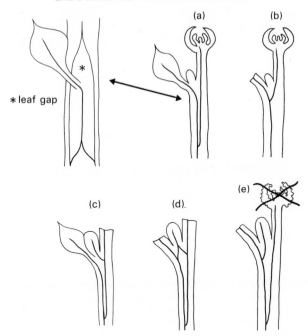

Figure 7.8 Experimental manipulation of vascular strand fusion in pea stems. (a) The vascular arrangement at an intact node. The vascular bundles differentiating from the apical bud and the young leaf avoid each other and only fuse further down the stem, forming a leaf gap in the vascular system. (b–e) When a leaf, or axillary or apical bud, is excised its strand can then be joined by other strands that still have growing tissues and presumed auxin sources at their tips. See text for further explanation. (After Sachs 1968)

They also indicate that a vascular strand having a source of induction at its distal end inhibits newer strands from fusing with it. Since apical dominance can be restored by replacing the excised apical bud with applied auxin, it was a logical next step to investigate the effect of applied auxin sources on the induction and differentiation of vascular strands.

Pea seedlings were decapitated and the vascular tissues were pulled out from the **epicotyl** stump. Auxin (IAA) applied in lanolin was able to induce a new strand of vascular tissue (Fig. 7.9a). If two similar auxin sources were applied simultaneously, two parallel strands were formed. However, if a strand was induced by auxin and then the auxin was removed and a second strand was induced by a second auxin source, then the second strand joined the first (Fig. 7.9b). The conclusion was that a vascular strand having a source of induction present was not joined by a second induced strand. Only when the source of induction of the first strand was removed would this allow a second strand to join it. These were the same conclusions as were drawn from the experiments involving natural inducing sources and the excision of the apical bud and

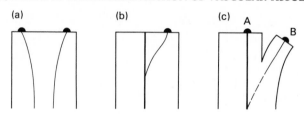

(d) **% of plants showing A – B contacts**

IAA at A	IAA at B		
	0.03%	0.1%	1.0%
0%	77	100	100
0.03%	33	71	100
0.1 %	43	57	87
1.0 %	4	13	18

Figure 7.9 Induction of new vascular strands by auxin (IAA). (a) Auxin can induce a vascular strand in pea epicotyl tissue. Twin auxin sources induce parallel strands that do not fuse. (b) An auxin source induces a strand that fuses with the existing strand of the epicotyl if this does not have an auxin source. (c) Auxin was applied at A to the epicotyl vascular strand. A new strand was induced by auxin applied at B. (d) The frequency of fusion of the B strand with the main strand increased as the concentration of auxin at B relative to A was increased. (After Sachs 1969)

young leaves (Fig. 7.8). The present experiments showed that auxin could act as the inducing agent and that it could mimic the effect of the apical bud, a growing axillary bud, or a leaf.

The next question was whether the relative strengths of inducing sources could affect the probability of two vascular strands joining. The epicotyl was cut to provide a flap of cortex on which auxin could be applied and form a strand that, in its upper part, would be separate from the main vascular bundle (Fig. 7.9c). It was found that whether the newly-induced strand at B joined with the pre-existing strand depended on the relative strength of the auxin source at B compared with that at A, on the epicotyl vascular strand (Fig. 7.9d).

These experiments lead to the following conclusions:

(1) Auxin by itself can induce vascular strands and may replace a bud or a leaf in this respect.
(2) The presence of auxin in a strand inhibits other strands from fusing with it.
(3) The relative amounts of auxin present in adjacent strands or flowing through them determines whether or not they will fuse.

The pattern of the vascular tissue in the stem, as shown especially in

Coleus by Bruck and Paolillo, can be accounted for by the production of auxin by the young leaves and a decreasing effectiveness of the leaves as inducers as they age. In all these experiments, vascular strands were identified by the xylem elements, which are the most easily seen in cleared tissue, but there is no reason to believe that phloem was not formed too. These experiments show again that auxin can be a critical and controlling factor in the differentiation of vascular tissues.

7.6.1 The course of vascular strand induction

Whan vascular strands are induced they differentiate basipetally, towards the base of the stem. To find out how vascular strand formation is controlled, T. Sachs experimentally manipulated the position and direction of strands in bean hypocotyls (Figs 7.10 a–n). After each experiment, the tissues were cleared so that existing and new xylem vessels could be seen and recorded. When a piece was cut out of the side of the hypocotyl, vessels were differentiated round it (Fig. 7.10a). If the piece was not cut out entirely but was left attached at its base (Fig. 7.10b) or at its top (Fig. 7.10c), vessels did not differentiate up into it (Fig. 7.10b) but they did differentiate down (Fig. 7.10c), ending blindly when they encountered the cut surface. When vessel strands were induced with auxin, they differentiated only basipetally (Fig. 7.10d) even if they ended blindly (Fig. 7.10e). The stimulus for differentiation could pass across a graft (Fig. 7.10f), but where the lower tissue had been morphologically inverted, the vessels in it differentiated upwards on the lower side of the graft (Fig. 7.10g) so that they were still differentiating in a morphologically downward direction.

So far, these experiments show that vessel differentiation is basipetal, i.e. down the tissue axis. Vessels could also be made to differentiate transverse to the tissue axis, across the hypocotyl, when a bridge of tissue was almost isolated by cuts (Fig. 7.10j) or when auxin was supplied at the side of an isolated ledge of tissue (Fig. 7.10h). Vessels could even be made to differentiate upwards if this was forced by strategically placed cuts (Fig. 7.10k). As in the pea, auxin alone could induce strands (Fig. 7.10l), and when there were competing sources of auxin the first established or the stronger source prevented the strands induced by a second source from fusing with its own strands (Fig. 7.10i). Auxin could also induce strands having a transverse polarity for part of their length (Figs 7.10h & m).

The simplest explanations for these experiments are:

(1) The cells of hypocotyls, stems, and roots (the plant axis), are polarized basipetally and, in the intact plant, vascular strand differentiation and auxin flow are predominantly basipetal, according to this polarity.

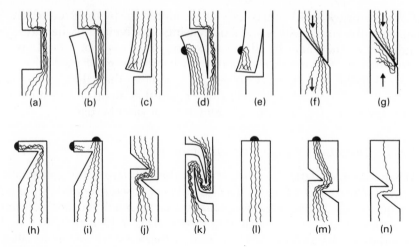

Figure 7.10 Polarity of vascular strand differentiation. (a–n) The effects of cuts made in the hypocotyl of *Phaseolus* seedlings on the differentiation of new vascular strands (wavy lines; original strands not shown). Auxin sources (1% IAA in lanolin) are shown as black blobs. Arrows show original direction to roots. The effects were not due to gravity, since horizontal plants gave the same results. Open-ended sections indicate continuity with the rest of the seedling. (After Sachs 1984)

(2) Auxin induces vascular strands.

(3) When auxin flow is forced to go across or against this polarity, a new polarity can be induced (Figs 7.10h & m), implying that a new polarity can be induced by auxin.

Box 7.1 Polar auxin transport[7]

Auxin may move around the plant in all directions in the vascular system. However, it is the only type of substance known to be transported from cell to cell in plant stems and roots in a polar fashion. This may help to explain its effectiveness as an inducer of strands of vascular tissue. Transport of auxin in the shoot-to-root direction (i.e. basipetal) is greater than in the reverse (acropetal) direction, irrespective of the orientation with respect to gravity. It continues even against an external concentration gradient. Basipetal polar auxin transport has been demonstrated in many plant stems and roots, which show a morphological polarity. In roots, the transport polarity is still in the morphologically downward direction, i.e. towards the root tip. Using [3H] IAA, it has been shown that the IAA moves as a wave down the tissue at a velocity of about 1 cm h^{-1}.

Experiments with metabolic inhibitors, such as dinitrophenol, showed that it was the secretion of auxin from the base of the tissue that was the active process, not its absorption at the top. Since the ratio of basipetal : acropetal movement increased with the length of the axis tested and, therefore, the more cells that the auxin passed through, this suggested that the transport mechanism involved secretion at the basal end of each cell. Naphthylphthalamic acid (NPA) is a potent inhibitor of auxin transport. When applied to sections of pea stem it was located by the use of fluorescently-labelled antibodies and shown to accumulate at the basal ends of the cells, implying that this is where the auxin channels are located.

The mechanism of IAA transport has been suggested as being first a diffusion into the cell of undissociated IAA (to which the plasma membrane is relatively permeable). The auxin outside the cell in the cell wall will tend to be undissociated because of the relatively low wall pH. Once in the cell, the IAA will tend to dissociate because of the higher intracellular pH. The IAA then tends to diffuse out of the cell at the basal end, where the auxin channels in the membrane appear to be concentrated. This outward diffusion is again down the diffusion gradient because of the lower concentration of IAA$^-$ outside the cell membrane. The polar movement of IAA is, therefore, thought to be a passive process down diffusion gradients. The system is maintained by outwardly directed proton pumps, which keep the cell wall at a lower pH than the cell cytoplasm. We do not yet know how this polar transport system originates, or whether auxin is involved in its establishment or simply makes use of a cell polarity produced by other processes. Not all substances showing auxin activity show polar transport. IAA (the main natural auxin) and NAA (naphthalene acetic acid, a synthetic auxin), are transported polarly but phenylacetic acid (the main auxin of peas) is not, nor are many synthetic auxins and auxin analogues.

7.6.2 Auxin flux and vascular differentiation[8]

Auxin flux is polar, the basipetal flux being several times greater than the acropetal flux (Box 7.1). Furthermore, in tissue showing polar auxin transport, auxin pretreatment can increase the polar auxin flux through that tissue. When vascular differentiation was induced round a wound in a stem, the basipetal flux of auxin round the wound had already increased after 16 h, although newly differentiated vascular tissue was not visible for three days (Fig. 7.11). Auxin, therefore, increases the flux of auxin through the cells while vascular differentiation is taking place. What happens when the axis of differentiation becomes orientated to become transverse (as in Fig. 7.10j)? Two cuts were made in a bean hypocotyl to force differentiation of transversely orientated vascular strands. After three days, the piece of tissue between the cuts was excised and the transport of auxin across it was measured with [³H]IAA. The *transverse* rate of auxin movement, across the tissue in the shoot to root direction, was twice that in the opposite direction (Fig. 7.12). Since auxin alone could induce similar transverse strands in a

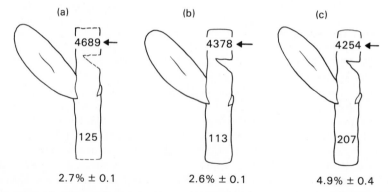

Figure 7.11 Auxin flux around a wound before differentiation of vascular tissue. A wedge-shaped wound was made in sections of bean stem bearing one cotyledon (to supply nutrients). [³H] IAA was applied where shown by the arrows. Numbers are radioactivity counts per minute above and below the cotyledon and percentages of auxin transported to the lower part. Stem cut from seedling (a) just before auxin applied, (b) 16 h before, and (c) 16 h before but with non-radioactive auxin present until its replacement after 16 h by [³H] IAA. Auxin pretreatment (c) gave a small but repeatable increase in auxin transported. (Sachs 1981)

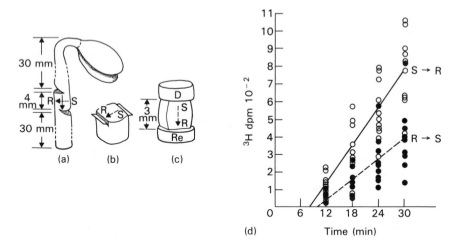

Figure 7.12 Induction of a transverse polarity of auxin flux by inducing transverse polarity of vascular differentiation. (a) A transverse bridge was made by cuts in a *Phaseolus* bean hypocotyl. (b) After 3 days the piece of hypocotyl was excised and the sides of the tissue were trimmed off. (c) The rate of *transverse* auxin movement in each direction across the hypocotyl was measured (S→R; shoot to root direction, and the reverse). The piece of hypocotyl was placed between donor (D) and receptor (Re) blocks of agar. (d) [³H] IAA in the donor block was transported twice as fast across the hypocotyl tissue in the S→R than in the opposite (R→S) direction. (Gersani & Sachs 1984)

decapitated plant (Fig. 7.10m), these experiments suggest the following conclusions:

(1) Auxin can induce vascular strands and, as it does so, increases the rate of polar auxin transport.
(2) Transverse reorientation of polar auxin movement and vascular strand differentiation are associated and both have the same polarity.
(3) The direction of polar auxin flow seems to determine the polarity of vascular differentiation.

Especially in dicotyledonous leaves, there is a network of veins so that there must be many stretches of vascular tissue that have either no polarity or dual polarity if, for instance, different individual vessels in a strand are polar but in different directions. Attempts to assign polarity to strands are bound to produce anomalies and contradictions (Fig. 7.13). Experiments were, therefore, carried out by T. Sachs to find out whether similar non-polar vascular strands could be induced by regularly altering the direction of auxin flow.

A pea seedling was decapitated and a horizontal wound was made in the epicotyl stump. When auxin was applied to one side of the wound, or both sides simultaneously, the xylem strands that differentiated were polar and the strands did not fuse, in agreement with previous experiments. Auxin could also be applied to one side of the wound at a time (Fig. 7.14). Each day for the next four days, the auxin (in lanolin) was removed and a new auxin blob was placed on the opposite side of the wound. So, for five days the auxin source alternated between the opposite sides of the wound. This caused transverse non-polar xylem strands to form above the wound in many of the plants. Simply replacing the auxin source at the same side of the wound did not have this effect – polar strands were formed instead. The formation of the non-polar xylem strand was, therefore, the result of the alternation of the position of the auxin source and the direction of auxin flow.

This experiment shows that auxin can induce a transverse axis of vascular differentiation without necessarily inducing an overall polarity. For the induction of a polar strand of xylem, other experiments showed that auxin had to be present for the whole of the three-day period necessary for there to be visible differentiation. During the first two days, the potential strand could be diverted by a second auxin source (as in Fig. 7.9a) and so its cells had not yet become determined. Auxin, therefore, seems to be able to determine the formation of an axis or strand of differentiated vascular cells but the polarity seems to be determined by the direction of flow of the auxin. This is normally determined by the existing polarity of the cells, which transport auxin basipetally. Whether

157

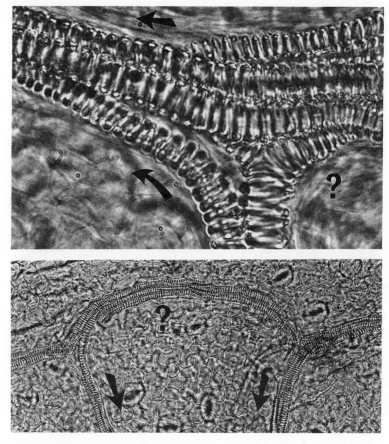

Figure 7.13 Polarity in vascular networks. Where vascular strands have closed loops, as in leaves, some vessels in some strands cannot have an unambiguous polarity. (Sachs 1981; photographs kindly supplied by Professor T. Sachs)

this basipetal polarity is originally induced by auxin during embryogenesis is not clear (see Ch. 10, section 10.1).

Since auxin can increase the auxin transport capacity of the cells and since the direction of auxin flow seems to determine the polarity when vascular strands are reorientated, it may be concluded that it is the flow of auxin through the cells and its direction, rather than its local concentration, that causes differentiation of vascular strands. Vascular strands can be induced across grafts (see Fig. 7.10f) into regions of pea stem cortex where there were originally no strands. This shows that:

(1) the inductive stimulus can pass across a graft;
(2) the stimulus presumably flows through the existing vascular strands to the graft surface;

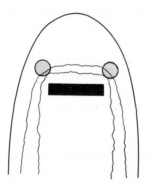

Figure 7.14 Induction of non-polar vascular strands by auxin. The epicotyl stump of a decapitated pea seedling was cut longitudinally and obliquely. A small transverse wound was made near the tip of this cut epicotyl. An auxin source (0.2% IAA in lanolin) was applied at one side of the wound and each day was removed and replaced with an auxin source at the other side, at the shaded sites. Continuous vascular strands (wavy lines) were often formed, which at some point have no polarity relative to the roots. (Sachs 1981)

(3)   new strands are induced to be continuous with existing strands.

Taken in conjunction with the effects of auxin in inducing strands and increasing auxin transport within them, these results imply that the newly induced cells themselves act as conduits for the stimulus and that, as they differentiate, their capacity as carriers of the stimulus improves and so promotes the induction of the cells beyond them.

It may not be the concentration of stimulus or auxin within the differentiating strand that matters. In pea roots, differences in the concentration of auxin supplied as inducer alter only the width of the induced strands; 0.04% auxin induces a narrow strand, and 0.2% a wide one. But the width of each strand is constant down through the tissues. If the local concentration of auxin changed down the strand then we might expect to see strands wider at the auxin source and diminishing in width with distance from the source, but we do not. The conclusion is, therefore, that the auxin moves through the differentiating strand but its concentration remains unchanged, and it is the directional flow of auxin through the cells that is critical. Support for this idea comes from experiments with okra stem sections. A similar concentration of IAA could be produced in the middle of the sections by basipetal or by acropetal transport, according to which end of the section IAA was applied. Only basipetally transported IAA caused differentiation of appreciable numbers of tracheary elements around a wound (186 compared with 16 for acropetally transported IAA).

Vascular strands (even those that are non-polar overall) probably develop only when there is polar auxin movement or in already polarized

159

tissues. If the polarized movement of auxin or stimulus results in a local concentration at the tip of the differentiating strand, then it may still be that such a local concentration has to reach a critical level to initiate differentiation but that a lower concentration can sustain the differentiation process. It would be extremely useful to have techniques that would enable us to know how the natural concentrations of auxin did or did not change within a differentiating strand and within its component cells. The amounts of auxin applied experimentally are usually so high that they can be of little or no value in simulating what actually happens in the plant. We also need to know whether auxin is confined to developing strands. Does it leak out? If not, why not? If it does, why is vascular differentiation not more diffuse? Clearly, there is room for many more experiments.

Vascular strands have been induced only in tissues that normally can give rise to vascular tissues. They tend to form most easily in the region of the cambium or procambium or from those cells of the stem that appear to be pith but are actually interfascicular (between existing vascular bundles). They also form most easily near wounds, the wounding apparently conferring on the neighbouring parenchyma cells the competence to differentiate as vascular tissue.

7.7   SECONDARY VASCULAR TISSUES[9]

Secondary thickening to produce an increase in girth of trunks, stems and roots increases the total amount of vascular tissue to allow the continued growth of woody plants, trees, and shrubs. This lateral growth is the result of the activity of the cambium, which is a cylindrical meristem lying at the inner surface of the bark. The cambium divides to produce phloem cells to the outside and xylem cells to the inside. Interfascicular cambium forms between the vascular bundles from parenchymatous cells. Although these cells may appear to be undifferentiated parenchyma, they may in fact already be determined as procambium. This is shown by Siebers' experiments in which they preserved their polarity when isolated and cultured or when they were excised and replaced in the stem but in reverse orientation. In the intact plant, the activity of the cambium and the differentiation of its products may be controlled by stimuli from the shoot apex and the leaves. The formation of vascular tissues by the cambium entails, first, the initiation and maintenance of cambial growth and division and, second, the control of the differentiation of the cells that it produces.

7.7.1   Cell division in the cambium[10]

There is considerable evidence that in deciduous trees auxin increases in a basipetal wave down the tree in the Spring and that cambial reactivation after winter dormancy is also basipetal. Auxin applied to decapitated plants can stimulate cambial activity. However, when examined in detail it has been found that cambial reactivation may occur just before the auxin wave reaches the cells and before the natural endogenous auxin concentrations increase. Also, in islands of bark isolated by a groove all round them, cambial activation in the Spring still takes place even though these isolated cambia could not have received any stimulus from above via the phloem or cambium. Differentiation of these cambial derivatives is poor, though, and the cambial activity soon peters out. These experiments suggest that cambium activation may not depend on auxin, but happens at about the same time that auxin begins to be produced by the buds that are also coming out of dormancy. Cambium that has not been released from dormancy is unresponsive to added auxin, and even cambium that is not dormant may not always respond.

Experiments in which leaves are removed from growing plants and are substituted by various growth substances suggest that leaves supply growth substances that are necessary for the sustained action of the cambium and also for the differentiation of its derivatives. When a *Xanthium* plant is decapitated, cambial activity continues at one-third of the control rate but the cells that it produces do not differentiate. Differentiation can be restored by a single, rapidly growing leaf or by the application of auxin and gibberellin (GA). Gibberellin applied alone increases the production of cambial derivatives but they do not differentiate. Auxin promotes the differentiation of xylem fibres. The auxin NAA was effective but IAA was not. This recalls the similar inability of IAA to induce fibre differentiation in peas (see section 7.6). Gibberellin similarly promoted the production of cells but not their differentiation in *Robinia*, whereas auxin (IAA) promoted differentiation of both xylem and phloem. When phloem differentiation was looked for carefully with the aid of polarized light, it was found to be stimulated by GA_3 in *Pinus*. In other species treated with GA, even if some phloem differentiation were found by careful examination, it would still be seen less easily and be less complete than in intact plants. In *Eucalyptus*, GA applications were shown to accelerate xylem production without phloem production being affected. This was shown by briefly supplying $^{14}CO_2$, which labelled a band of xylem and of phloem cells that provided reference points as new unlabelled cells were formed. In *Picea*, application of the cytokinin, benzylaminopurine, caused the production of large multiseriate **rays** so that the proportion of ray tissue in the xylem was increased.

The type of xylem cells formed in bean (*Phaseolus*) stems could be controlled by selectively cutting off leaves of different ages. The young leaves promoted the formation of **xylem vessels**, and old leaves promoted the formation of xylem fibres. When excised leaves were replaced by applied growth substances, it was shown that IAA could substitute for a young leaf in forming xylem with about 20% vessels. Gibberellic acid was required in addition to IAA to replace an older leaf, both in terms of the amount of xylem formed and the type of xylem, i.e. mainly fibres and poor in vessels. Abscisic acid (ABA) greatly reduced the amount of xylem formed but increased the proportion of vessels in it.

Growth substances, therefore, can control the activity of the cambium and the amount and type of secondary tissues formed, and can mimic some of the effects of leaves. Whether growth substances are in fact the actual controlling agents in intact plants is questionable. Their effects may differ from plant to plant and from season to season, and undoubtedly depend on their interaction with other metabolites and the responsiveness of the tissues to them.

7.8 SUMMARY

(1) Cell differentiation in plants is characterized by modification of cell shape, cell wall structure, and lysis or partial lysis of cell contents, as well as by the nature and composition of the cell contents. There are about 40 distinguishable cell types in plants, compared to 100 or more in animals.

(2) When a vascular bundle is severed, regeneration of vascular strands round the wound is mainly polar and basipetal, and is promoted by a leaf above the wound or by substituting the leaf by auxin. In isolated internodes, vascular regeneration may also require a supply of cytokinins, carbohydrates, and other substances otherwise supplied by the rest of the plant.

(3) Differentiation of xylem tracheary elements in tissue cultures and mesophyll cell suspensions does not generally require cell division or DNA synthesis, except possibly for a residual requirement for DNA repair or the synthesis of a minor DNA component. Cells can differentiate from either the G_1 or G_2 phase of the cell cycle. Cell division may still be required for early stages of cell differentiation.

(4) Vascular tissues can be induced in undifferentiated callus by the application of sucrose and auxin only. The tissues differentiate in normal orientation, with phloem outermost and xylem innermost. Auxin concentration seems to control the position of differentiation, and sugar concentration to control the types of cell formed (xylem, phloem, or cambium) and their proportions.

(5) Phloem can form alone, probably always preceded by cell division,

but xylem formation in the intact plant always seems to require phloem formation first. Phloem may also require lower concentrations of auxin for initiation than xylem. Cells in suspension can transform directly into xylem elements.

(6) Vascular strands are induced by apical and axillary buds, young leaves, and auxin sources. Auxin alone will induce vascular strands in polarized tissues. The fusion of vascular strands probably depends on the relative flux of auxin through them. Strands with a presumed high auxin flux prevent strands with a smaller flux from fusing with them.

(7) Vascular strands with a new polarity can be induced by restricting the pathway for basipetal auxin movement. Transverse polarity can also be induced by an auxin source placed on one side of a section of stem tissue. A new non-polar axis can be induced by alternating the auxin source daily from one side of the tissue to the other. Auxin not only induces polar vascular strands but also increases the auxin flux in the differentiating strand as it does so.

(8) Secondary vascular tissues are formed by the action of the cambium, which becomes activated in the Spring, after dormancy, at about the same time as a wave of auxin passes down the stem from the buds above, also coming out of dormancy. The amount and type of vascular tissue formed by the cambium can be regulated by auxins, gibberellins, and abscisins, which can be substituted for excised leaves and restore differentiation to resemble that found in the intact plant.

FURTHER READING

Aloni R. 1987. Differentiation of vascular tissues. *Annual Review of Plant Physiology* **38**, 179–204.

Barnett J. R. (ed.) 1981. *Xylem cell development*. Tunbridge Wells: Castle House Publications.

Bengochea, T., G. I. Harry, J. H. Dodds, R. Phillips, S. M. Arnott & R. A. Savidge 1983. Four papers forming a mini-symposium on differentiation in tissue cultures and the role of hormones in vascular differentiation. *Histochemical Journal* **15**, 411–18, 427–66.

Jacobs W. P. 1979. *Plant hormones and plant development*. Cambridge: Cambridge University Press. (An individualistic view, including vascular regeneration around wounds)

Roberts L. W. 1976. *Cytodifferentiation in plants; xylogenesis as a model system*. Cambridge: Cambridge University Press. (Earlier work, with emphasis on effects of growth substances)

Sachs T. 1981. The control of patterned differentiation of vascular tissues. *Advances in Botanical Research* **9**, 151–262. (Summary and synthesis of work on induction of vascular strands)

Shininger T. L. 1979. The control of vascular development. *Annual Review of Plant Physiology* **30**, 313–37.

Sugiyama, M. & A. Komamine 1987. Relationship between DNA synthesis and cytodifferentiation to tracheary elements. *Oxford Surveys of Plants Molecular and Cell Biology* **4**, 343–6. (Necessity for some DNA synthesis in cell suspensions. Also references to main papers on *Zinnia* cell suspensions)

NOTES

1 Houck & LaMotte (1977). (Phloem regeneration, and cytokinin requirement)
2 Fukuda & Komamine (1980, 1981), Hardham & McCully (1982), Phillips (1981), Sugiyama *et al.* (1986). (Xylem tracheary elements)
3 Wetmore & Rier (1963). (Formation of vascular nodules in callus)
4 DeMaggio (1972), Jeffs & Northcote (1967), Rier & Beslow (1967). (Auxin and sugar as inducers of vascular tissues)
5 Thompson (1967). (Xylem differentiates on same side of barrier as phloem)
6 Bruck & Paolillo (1984), Sachs (1968, 1969), Young (1954). (Auxin replacement of excised leaves and as an inducer of vascular strands)
7 Goldsmith (1977), Jacobs & Gilbert (1983). (Polar auxin transport)
8 Gersani & Sachs (1984), Thompson (1970). (Induction of polarity of auxin transport)
9 Siebers (1971). (Polarity of cambium)
10 Evert & Kozlowski (1967), Philipson & Coutts (1980), Savidge & Wareing (1981), Waisel *et al.* (1966). (Control of differentiation of secondary vascular tissues)

CHAPTER EIGHT

Cell enlargement, maturation, and differentiation

8.1 LIMITS OF CELL DIVISION[1]

The root and shoot apical meristems are the source of all the cells for the continued extension of the plant body. Meristems grow away from the more basal cells that they have produced but, when measuring their growth, it is more convenient to use the meristem as the point of reference rather than some arbitrary point in the basal non-growing regions. We can then regard the cells as being displaced away from the growing tip by the growth of the cells of the apical meristem more distal to them. As they are displaced, the cells differentiate into the tissue systems of epidermis, cortex, and stele (and root cap in the root), and into the different cell types – xylem, phloem, fibres, pith, etc. As cells pass out of the meristem they stop dividing, they enlarge and vacuolate, and then stop growing. This cell growth is common to all cell and tissue types. It is accompanied by a loss of meristematic activity and ends with cessation of growth, and has been called cell maturation. This emphasizes that, although it may be regarded as a generalized differentiation of cells behind the meristem, it is a process that is superimposed on the differentiation of cell and tissues types.

As the cells are displaced backward through the meristem, they divide four or five times and the length of the cell cycle does not change much. At the basal, proximal end of the meristem, cell division stops. This happens not by a slowing down of the cell cycle but by cells dropping out of cycle and arresting in G_1 or G_2 (the pre- or post-DNA synthesis parts of interphase, respectively). The proportions of cells in G_1 and G_2 in the maturing region of the root are characteristic of the species. In the root cap, the cells soon stop dividing after a couple of divisions or so, and the cells differentiate rapidly as they are displaced to the outside of the cap and then sloughed off.

In order to explain the cessation of cell division at the proximal

165

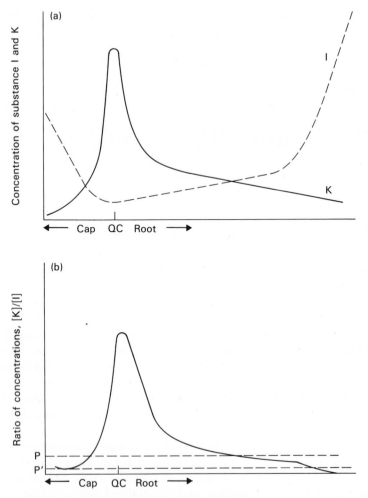

Figure 8.1 Model for control of cell division and enlargement in the root. (a) There are assumed to be two growth substances: K, which is synthesized in the quiescent centre (QC); and I, which is synthesized in the maturing cells of the cap and the root. At a K/I ratio higher than P, all mitotic events can go on; at a K/I ratio lower than P, cytokinesis is prevented and so endomitosis results in polyploidy; at a K/I ratio below P', no mitotic events can continue. Cells in those regions of the root tip with K/I ratios <P' are therefore non-dividing, enlarging, and maturing. There is insufficient experimental evidence either to accept or to reject this model. (Barlow 1976)

boundary of the meristem, it has been proposed that there could be two growth substances interacting to regulate the rate and extent of cell division (Fig. 8.1). The idea that substance K might be cytokinin seems less likely since it has been shown by immunocytochemical techniques that cytokinin concentration is low in the quiescent centre (QC). Abscisic acid (ABA), as well as auxin, may also increase in concentration basipet-

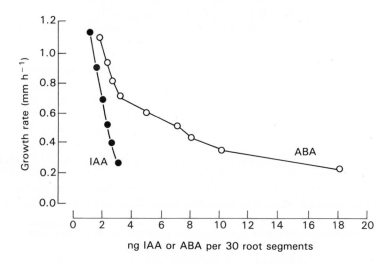

Figure 8.2 Growth rate of maize roots in relation to IAA (indole acetic acid, auxin) and ABA (abscisic acid) content. The fastest growing roots (measured over an 8-h period) had the lowest content of IAA and ABA in the elongation zone (segments cut 2.5 – 5 mm from the root tip). (After Pilet & Barlow 1987)

ally along the root and both may be correlated inversely with root growth rate (Fig. 8.2). The cessation of division in maturing cells may be the result of other metabolic changes in these cells (see Box 8.3). It may also be the result of decreasing cellular contacts. In *Azolla*, the frequency of plasmodesmata decreases, especially in the xylem and phloem, as the cells mature (see section 8.3.3).

8.2   CELL ENLARGEMENT AND MATURATION[2]

The maturation of cells can be studied most easily in the root because this grows essentially in one dimension, i.e. length. The spatial sequence of cells from the meristem to the mature part of the root, i.e. distance back from the meristem, represents a sequence in time. By comparing cells at successively greater distances from the promeristem, it is possible to follow the growth and maturation of an average cell (Box 8.1).

Box 8.1 Analysis of root growth[3]

(1) The root is cut into sections, each (usually) 1 mm long. By measuring
the number of cells per section as well as the amounts of protein, etc.
per section, the change in the average cellular composition with
increasing distance from the root tip can be found (Fig. 8.3). Because
the time for a cell to be displaced from one section to the next varies
continuously throughout the enlarging region (see point 2 below),
distance does not equate with a linear time scale. A necessary
assumption is that, whatever the age or length of the root, cells always
reach the same stage of development at a given distance from the root
tip. Although this is an approximation, since seedling roots change in
their growth rates and characteristics as they grow, it is sufficiently
close for most purposes. This method compares different cells with
different histories at a single point in time. Ideally, the same cells

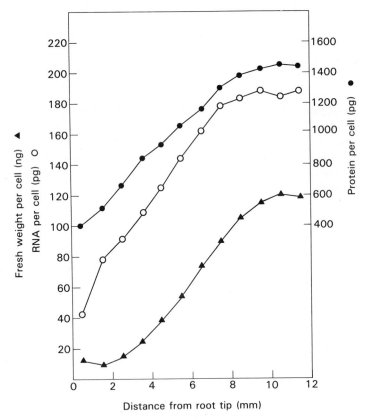

Figure 8.3 Increase in protein and RNA during cell enlargement in the pea root. The
cells are smallest in the meristem, and enlarge slightly in the root cap (at the tip) and
enormously in the root behind the meristem. The cells are mature about 10 mm from
the root tip. (After Heyes & Brown 1965)

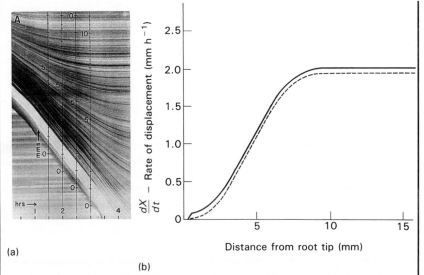

(a)

(b)

Figure 8.4 Cell displacement in the maize root tip. (a) Roots marked with lamp black were photographed continuously through a slit. The lines are points on the root surface (root tip at 0), traced as they are displaced downward with time. (b) From the trace is calculated the changing rate of displacement of cells as they pass from the meristem and through the enlarging region of the root. (Erickson & Goddard 1951; photograph kindly supplied by Professor R. O. Erickson)

should be compared at different elapsed times during their growth. This has been done with packet analysis (see point 3 below).

(2) Root cell growth can be put on a true time scale if we know the rates of cell displacement along the root and the time for displacement from one section to the next. Erickson and colleagues marked roots with lamp black and then made a continuous photographic record as they grew downward in a humid chamber. From the photographic trace (Fig. 8.4a), the rates of displacement of the marks (Fig. 8.4b) and the relative (elemental) rates of cell elongation (Fig. 8.5) were calculated. From counts of cell numbers in macerated sections, the rate of cell production was also calculated (Fig. 8.5). It is clear that in the meristem (terminal 2–3 mm), cell division and cell elongation are superimposed processes and are not mutually exclusive.

(3) In a root growing at a constant rate, the number of cells (n) displaced out of the meristem per hour will also be constant and the same number of cells (n) will be displaced per hour past any point in the enlarging region. Therefore, n cell lengths will represent one hour's growth. A graph of mean cell length as a function of number of cell lengths behind the meristem therefore describes the change in mean cell length per unit time. If the rate of growth of the root (mm h^{-1}) and the number of cell lengths added per millimetre in the mature region are known, then the abscissa can be made an absolute time scale.

(4) Packet analysis. Because the cell wall continues growing as the cell enlarges, groups of daughter cells can be distinguished because they are surrounded by the relatively thick wall of the original mother cell. Successive divisions produce new cell walls within the mother cell, the newest walls being the thinnest, and so the relative ages of the cells in a mother cell 'packet' can be deduced. Because the packets remain sufficiently distinguishable from each other as the root grows, the progress of packets of cells from the meristem to maturity can also be deduced. When combined with measurements of the overall rate of elongation of the root, and the frequency of number of daughter cells per packet, the rates of division and elongation of the cells can be worked out. By comparing roots at successive times during their growth, a true record of the cells' development can be constructed. Cell cycle lengths obtained by this method were comparable with those obtained by radioactive labelling (see Ch. 2, Table 2.1), which, as Green has pointed out, otherwise need to be accepted with caution because different cells are usually being measured at the start and end of labelling experiments. The agreement of cell growth values derived from packet analysis and other techniques shows that, despite their theoretical weaknesses, most methods in practice give acceptable results.

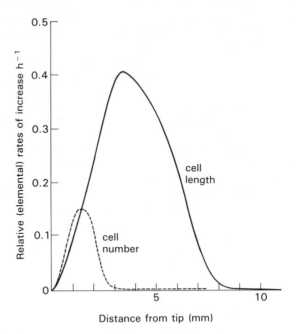

Figure 8.5 Relative rates of cell production and cell elongation in maize roots. The maximum rate of increase in cell number is near the root tip (in the meristem) but maximum elongation rate is in the enlarging region behind this. (Erickson & Goddard 1951)

Table 8.1 Changes in respiratory pathways during cell maturation in pea roots. Pea root tips were supplied with $[1-^{14}C]$ glucose or $[6-^{14}C]$ glucose. The expected ratio of C1 : C6 in the CO_2 respired is 1 if the glycolysis pathway is operating and >1 for the pentose phosphate pathway, which therefore becomes more important in the maturing cells. The activities of three enzymes of the glycolytic pathway decreased as the cells matured (PPK = phosphofructokinase; FDPA = fructose-1, 6-diphosphate aldolase; GPD = glyceraldehyde-3-phosphate dehydrogenase). (After Fowler & ApRees 1970)

Root segment		C1 : C6 ratio	Enzyme activity (nmol substrate consumed min^{-1} mg protein^{-1})		
Tissue or process	Distance from tip (mm)				
			PPK	FDPA	GPD
root cap	0 – 0.4	1.18			
meristem	0.4 – 1.0	1.21	95	311	245
vacuolation	1.0 – 1.6	1.19			
elongation	1.6 – 3.0	1.18			
progressive maturation	3.0 – 6.0	1.86			
of primary xylem and	6.0 – 16.0	2.66	56	177	100
phloem	16.0 – 26.0	2.97			

The cells enlarge as they are displaced back from the root tip. Although cell elongation occurs in all cells, including those of the promeristem, it is very slow indeed in the QC but is very rapid in the elongating region of the root behind the meristem (Fig. 8.3 & 5). Cell elongation comes to a relatively abrupt halt about 10 mm or less from the root tip. Any cells in the dividing region of the root that did not divide at the same rate as their neighbours, or that dropped out of the cell cycle for one or more cycles, would inevitably become longer than their neighbours. This is because all the cells at any given distance back from the meristem will be elongating at the same rate, as they are all linked together by their cell walls, which do not slip past each other. In any one tissue, all the cells are similar in length. Unusually long cells are very rare, so this means that neighbouring cells in a tissue have all had cell cycles of similar duration and gone through the same number of cell divisions before final elongation.

The cells, as they enlarge, also synthesize and accumulate cell wall material, proteins, and nucleic acids (Fig. 8.3). These data are averages for all the different cell and tissue types at a given distance from the root tip. This is obviously a simplified picture because although all the cells elongate at the same rate they expand radially and tangentially at different rates in different tissues. Cell division ceases sooner in the central stele and mid-cortex than in the epidermis, so the longest cells are in the stele and the shortest often in the epidermis. The protein composition, as shown by immunological tests and by gel electrophoresis of root extracts, changes as cells mature and also differs from tissue to tissue. The changes seen so far seem mainly quantitative rather than qualitative. Some of the changes could be due not to changes within particular cell types but to changing proportions of the different cell types, each of which may have a distinctive, but relatively constant, structure and composition. Such an explanation may at least partly account for the change in respiratory pathways with cell maturation (Table 8.1). The pentose phosphate pathway increases in importance relative to glycolysis as the cells enlarge. This is associated with an increase in the proportion of vascular tissues in which the pentose phosphate pathway predominates.

In the shoot, the cells show similar changes to those in the root during enlargement. The shoot differs in that the region of rapid growth is much more extended than in the root, so that most of the apical bud consists of growing and dividing cells and cell elongation continues into the extending internodes (see Ch. 2, section 2.3.2). The control of cell elongation, especially by auxin, has been studied almost exclusively in shoot tissues. (Box 8.2).

Box 8.2   Mechanism of cell elongation[4]

Cells can enlarge and elongate only as fast as the growth of their cell walls allows. The rate of cell enlargement (dV/dt) depends on wall extensibility (m) and on the pressure exerted on the wall by turgor pressure (P) in excess of the wall yield threshold (Y), so that:

$$dV/dt = m(P - Y)$$

The chemical changes in wall structure that allow wall loosening and, therefore, an increase in m are not definitely known but almost certainly involve breaking and re-forming cross linkages between the wall polysaccharides and between them and the wall proteins. When these processes can no longer occur then wall growth ceases (Box 8.3). Turgor pressure (P) may be increased by the synthesis of solutes in the cell, so decreasing the osmotic potential and (at constant water potential) automatically increasing turgor pressure (P). Wall yield threshold (Y) is a property of the wall, depending on its structure.

Changes in wall extensibility (m) have been measured often by the extensiometer technique. A piece of methanol-killed tissue is extended at a constant rate in the extensiometer apparatus and the plastic extensibility of the walls is measured from the rate at which the tension in the wall increases. Plastic extensibility is not a direct measure of m but seems to be proportional to it, so this technique can be used to detect changes in m and their approximate magnitude. Treatments of living tissues can cause measurable changes in plastic extensibility measured in this way.

The only growth substance having a consistent effect in increasing wall extensibility (m) is auxin, and usually it is necessary to excise the tissue and so deprive it of its normal auxin source in order to show an effect. Auxin acts in two phases, which may overlap: the initial growth response (first 1–2 h) and the prolonged growth response (3–6 h and longer). Both may begin within 10–20 min of the application of auxin. During the initial phase, proton secretion from the cell out into the wall increases, the pH of the wall decreases, and wall loosening is promoted. The wall-loosening effect of H^+ ions can be mimicked by placing the tissues in low pH buffer (about pH 4) instead of auxin. Lowering wall pH could activate polysaccharide hydrolases since many of the enzymes present in cell walls have optima in the range pH 4–6. Wall hemicellulose xyloglucans become partially degraded when tissues are treated with IAA or a buffer of pH 4. Proton extrusion into the cell wall may initiate auxin-induced cell extension but its continuation may depend on the other effects of auxin on changing the expression of the genes and causing the synthesis of a new enzyme complement, so producing a new metabolic state in the cell (see Ch. 9, section 9.4.2). Another wall-loosening factor may be the removal of Ca^{2+} ions, which inhibit wall loosening. If auxin promotes Ca^{2+} removal from the wall by its uptake into the cell, it could promote cell extension in this way, but so far there is no experimental evidence for this proposed mechanism.

In the large coenocytic alga, *Nitella*, cell extension is not sensitive to auxin. Regions of proton efflux in the cell can be seen as yellow regions around the alga when it is immersed in a solution of the pH indicator, phenol red. By placing resin bead markers on the cell surface, it has been shown that almost all elongation growth is restricted to the regions of proton efflux. In *Nitella*, the inner cell wall microfibrils have a predominantly transverse orientation and the wall apparently grows by multi-net growth (see Ch. 4, section 4.2). Auxin seems to be most effective in promoting cell extension in cells of higher plants in which the walls have a polylamellate structure (see Ch. 4, section 4.2). Auxin may be needed in such cells in order to promote wall loosening for extension to occur.

The mechanism of cell extension in roots, where auxin often inhibits elongation, has been investigated much less than in shoot tissues. Note that in roots the main region of cell elongation is that region behind the tip, where there is an outward ion current that is mainly carried by protons (see Ch. 6, Fig. 6.3c). Whether auxin transported basipetally from the shoot (see Box 7.1) is responsible for this is not known, but it may be released in the enlarging region despite the overall negative correlation between growth rate and IAA content observed in maize roots (Fig. 8.2).

The cells in the stem that are most responsive to auxin are the epidermal cells, and it is epidermal cell extension that seems to determine the extension of the stem as a whole. In the inner tissues of coleoptiles, wall extensibility (m) does not seem to alter but the length to which the tissues would extend by water uptake after peeling seems to keep ahead of the increase in coleoptile extension, presumably by a continuous maintenance of turgor by synthesis of solutes. In stems, it is the pith cells that mainly seem to respond to gibberellins. The growth of stems (and roots?) may, therefore, depend on the cooperation of outer cells responsive to auxin, and the inner cells responsive to gibberellin.

8.2.1 Cell maturation: how is it controlled?[5]

Cell enlargement and maturation are not controlled by the position of the cells in the root because these processes can continue in isolation from the rest of the root. Pieces of root cut from the region 2–4 mm behind the root tip, where cell division has ceased and elongation rate is about maximal, can be cultured in solution containing nothing but 2% sucrose. The cells in these pieces of root continue to elongate as in the intact root but with no net synthesis of cellular constituents (except for cell wall material) because they lack a source of nitrogen. Despite no net increase in protein, enzyme activities in isolated sections still change as they do in the intact root (Table 8.2) and the normal cell types differentiate. Since neither the root tip nor base is present, the continued elongation and differentiation of the cells cannot be a function of their position in the root. These processes seem to be automatic and self-sustaining, dependent on oxidative metabolism, and are set in train when the cells leave the meristem. The implication is that the events of maturation progress steadily with time as though governed by a clock started in the cells when they left the meristem.

Table 8.2 Ratio of invertase : phosphatase enzyme activities in cells of the enlarging region of *Vicia faba* roots. The ratio of activities increases and then decreases in the same way in intact roots and isolated segments, although the absolute time scales differ. (Lyndon 1979)

Intact roots		Isolated 3–4 mm segments	
Distance from apex (mm)	Ratio	Hours after isolation	Ratio
3 – 4	5.6	0	4.4
4 – 5	8.4		
5 – 6	9.6	6	6.6
6 – 7	9.1		
7 – 8	8.3	12	1.5
8 – 9	7.0	24	3.4
9 – 10	7.1		
10 – 11	4.6	36	0.8
11 – 12	4.3	48	1.7

Box 8.3 Why does cell growth stop in maturing cells?[6]

Once cell extension has stopped it cannot be restarted except by treatments that cause the cells to begin dividing again. As extension ceases, the cell walls become more rigid and less extensible. This is probably because of increasing cross-linking of the wall components. Also, as the vascular tissues differentiate, the xylem becomes lignified and rigid and so may mechanically prevent the growth of adjacent cells. Cessation of wall extension may be for any or all of the following reasons:

(1) Increased cross-linking of cell wall polysaccharides.
(2) Increased amounts of the cell wall protein extensin. This hydroxy-proline-rich protein is characteristic of cell walls and increases in amount during cell maturation. When extensin synthesis is inhibited by hydroxyproline or α, α' dipyridyl, which prevents extensin synthesis by blocking hydroxylation of the proline residues in the precursor protein to hydroxyproline, then cell extension is promoted in cultured root sections. It is probably not the actual amount of extensin that matters for restricting further cell extension but the degree of its cross-linking with wall polysaccharides. Extensin cross-links apparently cannot be reversibly broken to allow growth. Extensin is incorporated throughout existing cell walls, not just in the newly synthesized inner layers, but it does not pass into the **middle lamella** and so does not pass into the walls of adjacent cells.
(3) Increased peroxidase activity. This would (a) convert ferulic to diferulic acid, which can then act as a hemicellulose cross-link; (b) generate H_2O_2, which would facilitate the oxidation of cinnamyl alcohols to free-radical lignin precursors; (c) promote cross-linking of wall proteins by phenolic compounds; and (d) oxidize free auxin, which may be required for cell extension.

Support for this idea comes from roots growing at different rates. In slowly growing roots, the differentiation of cells is much closer to the apical meristem than in fast-growing roots, as would be expected if differentiation of the cells depends on the time since they left the meristem rather than their position in the root. Further supporting evidence comes from roots with the meristem cut off, in which the initiation of lateral roots occurs at the expected time in the appropriate cells, showing that position with respect to the root tip seems irrelevant. Experiments with the root cap have shown that, here, the situation may be more complicated. In roots treated with colchicine, the root cap did not divide or produce new cells and, therefore, the cells did not change their positions relative to each other, yet they continued to become endopolyploid with time. However, in untreated roots but with the cap bisected, mucilage was formed by the cells on the surface irrespective of whether they had formerly been on the outside or the inside of the cap and, therefore, irrespective of the time since they were meristematic. The cytoplasmic characteristics of the cells as seen by electron microscopy were retained according to their original positions in the cap. Their cytoplasmic development was, therefore, suspended when displacement no longer occurred. In the root cap, it looks as though nuclear events may be time-determined but cytoplasmic development position-determined.

If the cells, on leaving the meristem, enter a phase of development that depends on some sort of timing mechanism, what is the nature of the clock and how is it set going? It could be of the egg-timer type, where sets of enzymes and proteins are synthesized (and perhaps broken down as well) each at different (but constant) rates, so that the relative protein composition of the cells continues to change. Thus, maturation could be programmed as different but constant rates of synthesis for different proteins, triggered as the cells leave the meristem and stopped perhaps when the metabolic network has changed to a state that no longer supports or allows growth, as perhaps happens in the cell wall (Box 8.3). This hypothesis implies no changes in relative gene activities during the maturation process. Alternatively, there could be a cascade mechanism whereby the first changes trigger changes in gene activity, which then alters the type of protein being synthesized. This would imply a continuously changing regulation of gene activity. By analogy with fruit ripening (see Ch. 9, section 9.2.2), which may be regarded as a continuation of cell maturation, we would expect genes to be regulated as a function of time during cell maturation, perhaps as the result of changes in the amount of some growth regulator in the cells as a function of time.

In roots, there may be a gradient of basipetally increasing concentration of auxin and abscisic acid (ABA) (Fig. 8.2) and, since increasing concentrations of each of these can inhibit root extension and ABA can decrease wall extensibility, they could be involved in promoting cell maturation. However, these are unlikely to be involved in isolated root segments unless they are produced as the cells differentiate. In shoots, auxin is likely to decrease basipetally and there is no evidence for gradients of ABA. Although these growth substances may modify cell maturation, they are unlikely to be major controlling factors unless cell enlargement and maturation are controlled in different ways in the root and the shoot.

Cell maturation is the norm in all plant parts of limited growth, such as leaves, internodes, floral organs, and fruits. In leaves, the cells mature first at the tip of the leaf and maturation progresses down the leaf (Fig. 8.6). In grass leaves, the cells at the leaf base remain meristematic and, because the leaf is linear in shape, it can be cut into successive

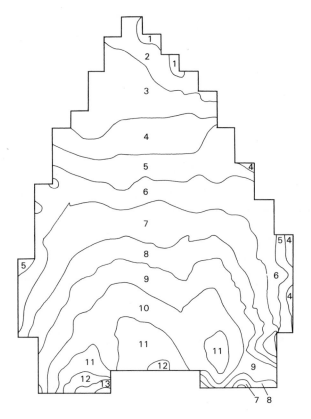

Figure 8.6 Relative area growth rates in a *Xanthium* leaf. Relative area growth rates increase down the leaf from $0-0.05\,\mathrm{d}^{-1}$ (position 1) to $0.05-0.10\,\mathrm{d}^{-1}$ (position 2) to ... $0.55-0.60\,\mathrm{d}^{-1}$ (position 12) and $0.60-0.65\,\mathrm{d}^{-1}$ (position 13). This is because the cells mature first at the tip of the leaf. (After Erickson 1966)

sections so that cell maturation can be followed as in the root. In wheat leaves, the development of the cells and the plastids has been followed as the cells mature (Fig. 8.7).

8.3   CELL DIFFERENTIATION

Plant cell types (see Ch. 7, Table 7.1) are distinguished by their morphological characteristics visible in the light microscope: cell size, shape, and position; cell wall structure; and cell contents (organelles, starch, pigments, and secondary products, such as crystals). Only wall structure and cell contents obviously depend directly on biochemical and molecular differences in the cells. Cell size may also reflect molecular differences: large cells, especially those of the developing xylem vessels, may

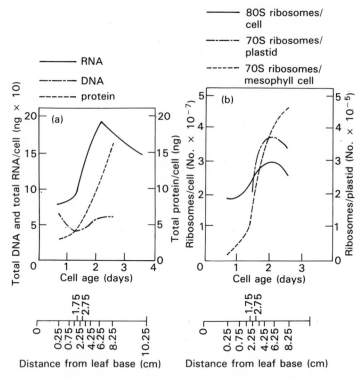

Figure 8.7 Development of wheat leaf cells. Sections cut from the leaf at increasing distances from its base represent a developmental time sequence. (a) As in the root, cell maturation is accompanied by increase in protein and RNA content. (b) Both cytoplasmic (80S) and plastid (70S) ribosomes are synthesized rapidly as the mesophyll cells expand. (Dean & Leech 1982)

be **polyploid**, and in the embryo, suspensor cells may be polytene, with increased amounts of DNA per cell and multiple genome copies.

8.3.1 Changes in cell wall structure

8.3.1.1 Localized cell wall synthesis[7]

The development of localized wall thickenings is especially characteristic of xylem elements. Xylem tracheids and vessels show a gradation of wall patterns, all of which are brought about by uneven thickening of the secondary cell wall. The main component of the thickened regions is cellulose. In *Zinnia* cell suspensions, immunofluorescent labelling has shown that the bands of wall thickening in developing xylem elements are preceded by a corresponding underlying pattern of microtubules (MTs) of the cytoskeleton (Fig. 8.8). Electron microscopy of other cells has shown the MTs lying parallel to the cellulose microfibrils in the

178

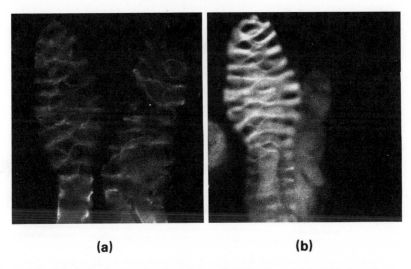

(a) **(b)**

Figure 8.8 Cytoskeletal patterning for xylem wall thickenings. (a) and (b) show the same two developing xylem elements in a *Zinnia* cell suspension. In the left-hand cell, the microtubule pattern shown by immunofluorescence (a) closely resembles the pattern of wall thickening shown by Calcofluor fluorescence (b). In the right-hand cell, the wall pattern has not yet developed (b) but the microtubule pattern is already there (a), showing that the cytoskeletal pattern precedes the wall pattern. (Roberts *et al.* 1985; photographs kindly supplied by Dr K. Roberts)

thickenings. When the MTs become bunched together to give the pattern, there is a concurrent rearrangement of actin filaments from longitudinal into transverse arrays. This suggests that the actin filaments may regulate the shift in the MT arrays but it leaves unanswered the question of how the actin filaments themselves become orientated. In addition to rearrangement of existing MTs to form the banded pattern, there is also synthesis of new tubulin to form new MTs. If the MTs are disrupted with colchicine, the thickening becomes irregularly plastered over the growing wall. When colchicine is removed, the MTs are reformed and the orderly wall thickening is resumed. These observations are consistent with the general view that wall microfibrils are somehow directed in their placement and orientation by MTs (see Ch. 4, section 4.2). In adjacent cells, the positions of thickenings often correspond, suggesting some sort of influence from one cell to the next (see Ch. 7, Fig. 7.2). Even when colchicine is present, some local wall thickening of xylem elements may still go on in that part of the wall next to existing thickenings in a neighbouring cell.

During xylem element differentiation in *Zinnia* cell suspensions, there is an increase in the activity of wall-bound peroxidase and PAL (phenylalanine ammonia lyase). These enzymes are necessary for the synthesis of **lignin**, which is incorporated into the xylem cell walls. However, lignin

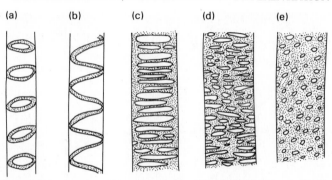

Figure 8.9 Xylem element structure. Wall thickenings range from (a) annular and (b) helical (typical of protoxylem) to (c) scalariform and (d) reticulate (metaxylem), and (e) pitted (often in secondary xylem).

synthesis is not essential for xylem differentiation. It simply gives the cells rigidity. If PAL is inhibited and lignin is prevented from forming, the cells collapse because they are less rigid. Lignin synthesis is, therefore, part of the process of xylem differentiation but its inhibition does not prevent other parts of the differentiation process from going on. Phenylalanine ammonia lyase activity is, therefore, a useful marker of xylem differentiation but its absence does not necessarily imply the absence of xylem differentiation. In bean callus, increases in PAL and callose synthase activity, associated with xylem and phloem differentiation, respectively, have been detected just before or during vascular differentiation.

In the intact plant, the thickenings of the xylem show a gradation from the annular and spiral thickenings of the protoxylem to the pitted thickenings of secondary xylem (Fig. 8.9). The development of complicated wall structures, such as bordered **pits**, is presumably in some way prescribed by the cytoplasm through the agency of the MTs, but the mechanism is as yet unknown.

Localized wall thickenings also develop as convoluted wall ingrowths in transfer cells. By increasing the surface area of the plasma membrane that has to cover these ingrowths, the area of the cell surface is thereby increased many-fold. Transfer cells are found especially in nodes and in regions near or in the vascular system (phloem companion cells sometimes develop as transfer cells), where high rates of solute transfer would be expected across the plasma membrane between symplast and **apoplast**. Transfer cells could be induced to develop in pea leaves in the dark by immersing them in 2% glucose solution for about 6 days. This suggests that, in the growing plant, transfer cells may develop in response to high local concentrations of sugars.

8.3.1.2 Localized cell wall breakdown[8]

In the xylem vessels, the end walls of the cells break down and disappear, and in the phloem sieve tubes, wall breakdown is again restricted to the end walls (sometimes parts of the side walls) but is only partial and ceases when the sieve pores have formed. Why are only the end walls affected by the lytic enzymes? The xylem end walls differ in composition from the side walls, being mainly pectin and hemicellulose with very little cellulose and no lignin. They can, therefore, be broken down selectively by the appropriate enzymes. The plasma membrane at the end walls may also be different in structure and properties from that on the side walls of the cells, as it is in the polarized *Fucus* zygote (see Ch. 6. section 6.1) and in auxin-transporting cells (Box 8.2). There are ion currents generated in growing root tips (see Ch. 6, Fig. 6.3) and these could conceivably be involved in the polarization of differentiation in developing vascular elements.

8.3.2 Cell contents

8.3.2.1 Changes in organelle content[9]

Among the first signs of cell differentiation in the meristems are changes in organellar composition of the cells. In the root apex, differences are mainly owing to cell maturation. In the shoot apex, organellar composition of the cells was shown to change according to how the cells would differentiate (Table 8.3). Cells forming the young leaf primordium had increased numbers of plastids and mitochondria. The incipient pith had no increase in plastid numbers per cell but the plastids enlarged as the cells enlarged and began to vacuolate. Since the number of plastids per cell increased in the incipient leaves, this means that plastid replication was faster than cell replication. In the incipient pith, plastid replication kept in step with cell replication because plastid number per cell remained constant but mitochondrial replication was faster. These examples show that cell division, and plastid and mitochondrial replication, can all be under separate control. How the rate of organelle replication is regulated is quite unknown.

8.3.2.2 Breakdown of cell contents

The most obvious case of lysis is the xylem in which mature cells have lost all their cell contents and are dead. Sclerenchyma fibres, too, are often dead. The phloem is curious because the lytic process stops half-way. The nucleus completely disintegrates in the sieve tube and the tonoplast is lost, so the cytoplasm and vacuole become mixed. The dictyosomes disappear. In the mitochondria, the internal membranes become much reduced and the endoplasmic reticulum becomes much more vesicular. Enzymes for callose synthesis presumably remain, since callose is formed

Table 8.3 Changes in organellar composition of pea shoot apex cells during early differentiation. In the young leaf primordium, plastid numbers increase. In the young pith, plastid numbers per cell have not changed but ER (endoplasmic reticulum) and vacuoles have increased and the cells have noticeably enlarged. At the site of the axillary bud, the cells, which remain relatively inactive, are smaller but similar in composition to the meristem cells. Pla, plastids; Mit, mitochondria; Dic, dictyosomes; ER, endoplasmic reticulum; Vac, vacuoles. (After Lyndon & Robertson 1976)

Cell type	Number of organelles per cell			Volume (μm^3) per cell		Cell volume (μm^3)
	Pla	Mit	Dic	ER	Vac	
promeristem	11	85	15	25	1	475
leaf primordium	16	85	23	34	2	491
axillary bud site	9	51	13	28	3	355
incipient pith	10	83	20	43	12	572

when the phloem is wounded. The characteristic phloem protein (P-protein) seems to increase in amount. The process of lysis itself implies the activity of lytic enzymes in the differentiating xylem and phloem cells. The formation of sieve pores has been followed with the microscope but, biochemically, little is known except that callose synthesis occurs. In arborescent monocotyledons, such as palm trees, which do not have a cambium, the phloem sieve tubes continue to function without a nucleus for the life of the tree.

8.3.3 Changes in cell-to-cell communication: plasmodesmata[10]

The cytoplasms of adjacent cells are linked through the cell walls by the plasmodesmata to form the symplast. Microinjection of fluorescent molecules of known sizes into cells shows that the plasmodesmata do not normally allow the passage of molecules bigger than about 850 daltons. But larger particles may be able to pass through the plasmodesmata, since some enzymes (e.g. RNase) and viruses can pass from cell to cell. The exclusion limit (or cut-off size for molecules) for movement through the plasmodesmata may possibly change during development, but whether this is important for development is not clear.

The number of plasmodesmata per cell wall may increase or decrease during development. In *Azolla* roots, there is a general loss in plasmodesmata as all types of cell develop and mature, but there is an accelerated loss in developing xylem elements. In the xylem cells, the plasmodesmata have vanished, by obliteration, by the time the secondary wall thickenings begin to be laid down. They are also lost in the developing phloem **sieve elements**, except for those few that remain in the transverse

182

walls and become transformed into sieve pores. All the cells in the *Azolla* root, therefore, become more and more isolated from each other symplastically as they develop, and this is accentuated for the xylem and phloem elements.

8.3.4 Biochemistry of xylem and phloem differentiation[11]

Differential extraction of onion root sections showed differences in cell wall composition during differentiation of the tissues. The cortical cells were relatively high in cellulose and low in pectin. In contrast, the vascular tissues were relatively rich in pectin and poor in cellulose, and later in their development hemicellulose became more abundant. The root cap differed from the rest of the root in having large amounts of cellulose and non-cellulosic polysaccharides.

The biochemistry of cell differentiation has been studied mainly in secondary xylem and phloem cells because these can be obtained in sufficient quantity at all stages of their development for their chemical composition and enzyme activities to be measured. In a cross section of the wood of a tree, the youngest and most recently formed xylem is at the cambium, and radial progression towards the stem centre follows the developmental progression from undifferentiated to fully differentiated xylem. A radial plug taken from a tree can be cut into sections, the distance from the cambium representing the sequence of development. Note that this is a study of differentiation in a tissue – xylem – that contains several different cell types.

As secondary xylem cells differentiate, the pectin content of the wall decreases and the xyloglucan content increases (Table 8.4). Pectin is characteristic of the primary wall. At the transition from primary to secondary wall formation, there is little or no change in pectin synthesis but a marked increase in the synthesis of hemicellulose and cellulose. This is reflected in the activities of the enzymes that interconvert the UDP-sugars to form the precursors for polysaccharide synthesis. The epimerases forming the pectin precursors were most active in the cambium and least active in the differentiated xylem, but enzyme activity could still be shown even when no pectin was being synthesized. These epimerases could not, therefore, be a major control on pectin synthesis.

The hemicelluloses in angiosperm xylem are formed almost entirely of xylans (formed from UDP-D-glucose and UDP-D-xylose), whereas in **gymnosperms** they have a high content of galactoglucomannans (formed from UDP-D-galactose and UDP-L-arabinose). The enzymes synthesizing UDP-D-glucose and UDP-D-xylose increase threefold during xylem differentiation in the angiosperms sycamore and poplar but, as expected, decrease in pine and fir. Although these epimerase enzymes are, in general, most active when they would be most required, they are active

Table 8.4 Changes in xylem cell wall composition during differentiation of sycamore wood. As the cells develop, they stop synthesizing pectin but become rich in cellulose and hemicellulose and become lignified. (After Northcote, 1963)

	Cambium (ng cell^{-1})		Sapwood (ng cell^{-1})	
pectic substances	2.3	(16%)	5.8	(4%)
α-cellulose	5.6	(38%)	62.8	(40%)
hemicellulose	6.9	(46%)	49.5	(31%)
lignin	–	–	39.7	(25%)

to some extent at all times. The activities of the epimerase enzymes do not, therefore, seem to be major controls of polysaccharide synthesis.

On the other hand, the enzymes synthesizing the sugars themselves show marked changes in activity. Polygalacturonic acid synthase, which is concerned with pectin synthesis, decreases as the cells develop and pectin synthesis slows down and stops. Conversely, the xylan content of the cells increases as they develop and xylan synthase activity increases 15 times in sycamore and 4 times in poplar. Phenylalanine ammonia lyase activity, necessary for lignin synthesis, increases ten times. Glucomannan synthase activity in pine increases fourfold. The changing activities of the synthase enzymes are much better correlated with the rates of synthesis of the wall components and are, therefore, better candidates as controlling agents in the synthesis of the cell wall in xylem differentiation.

Arabinan synthetase activity is induced in bean cell suspension cultures by transferring them to a medium in which they will divide and grow but not differentiate. When the cells are transferred to a medium with cytokinin, and xylogenesis is induced, xylan synthase activity is also induced. Both enzyme activities are inhibited by actinomycin D (an inhibitor of gene transcription) and also by MDMP (an inhibitor of translation of mRNA). Induction of these enzymes can, therefore, be brought about by changing the chemical composition of the medium and may be controlled at both transcriptional and translational levels.

8.3.5 Early stages of cell differentiation[12]

The study of cells *in situ* in sections by histochemical methods has two advantages. Individual cells and cell types can be studied, and the positions at which differentiation of particular cell types will occur in the meristems and young tissues can be predicted and so development may be followed from its earliest stages. Esterase activity can be visualized histochemically and is characteristic of vascular tissues, especially xylem. Using it as a marker for xylem, esterase activity in the root is traceable right back to the initials in the region of the quiescent centre. In the shoot, cell differentiation cannot be traced back to the initials in the way it can in the root. It is first visible below the apical dome, although esterase activity can be detected within the apical dome and may indicate early procambial differentiation as it does in the root. Procambium may also first be detected in the pea shoot apex as a local increase in the rate of cell division.

8.3.6   Nuclear changes during cell differentiation[13]

The most obvious nuclear changes are those in the developing xylem vessel elements. These cells are most easily seen in, for instance, maize roots where they can be traced back to the quiescent centre. Cell division ceases soon after the cells leave the initial zone but growth continues and so does DNA replication, in step with growth so that these cells become polyploid (Fig. 8.10), sometimes as much as $64n$. Each round of DNA replication is accompanied by an increase in ribosome content of the cells so that protein synthesis is maintained. This is a way by which the cells can replicate the genes, including rRNA genes, which seems to be necessary to allow the cells to enlarge so much. These cells may also provide a pointer to one possible reason why cells mature. If there is a maximum cell or cytoplasmic volume that can be supported by a given amount of DNA, then the cessation of cell division at the proximal edge of the meristem may then place a limit on subsequent growth.

The xylem vessel elements are not the only cells that may become polyploid as they mature. Individual cells scattered throughout the tissues, but especially in the cortex, may become polyploid. This is much less common in monocotyledons than in dicotyledons, and seems to be rarer in the shoot than in the root. So far, no role has been ascribed to polyploidization. It seems to be, in itself, a form of differentiation of individual cells superimposed on the differentiation of the tissue as a whole. Cell enlargement in the suspensor of some embryos, e.g. in the Leguminosae, is associated with DNA replication without cell division but in this case the cells either become multinucleate coenocytes or the chromosomes become polytene and closely resemble the giant chromosomes of *Drosophila* salivary glands.

When maturing cells stop dividing they may drop out of the cell cycle either in G_1 or G_2. Cells in different tissues in a root may tend to accumulate in different parts of interphase when mature. Different species show characteristically different proportions of cells in G_1 and G_2 (Fig. 8.11). The nutritional state of the root when the cells are maturing can also affect the proportions of cells arresting in G_1 and G_2. The significance, if any, of the mature cells becoming 'parked' in one part of interphase rather than another is not known.

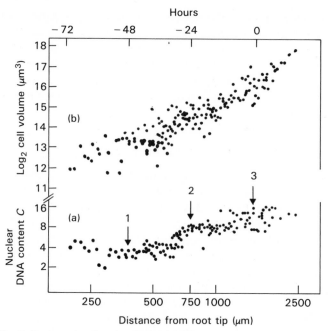

Figure 8.10 Enlargement and endopolyploidization of metaxylem cells in maize roots. (a) As the metaxylem elements grow, mitosis ceases (position 1), then the nuclei become tetraploid, 8C (position 2), and octoploid, 16C (position 3). (b) The steps of ploidy increase are less evident in the cell volumes. (Barlow 1985)

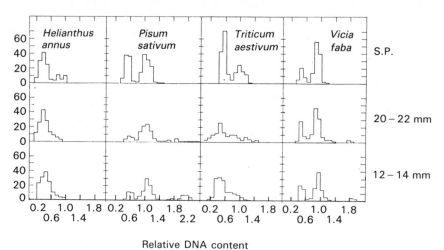

Relative DNA content

Figure 8.11 DNA content of nuclei in maturing pea root cells. Cells in the mature region of the root (12 – 14 mm and 20 – 22 mm behind the root tip) remain in either G_1 (relative DNA content 0.5) or G_2 (relative DNA content 1.0), the proportions of G_1 and G_2 cells being characteristic of the species. Meristem cells prevented from dividing by carbohydrate starvation (SP) also suspend growth with the same species-characteristic proportions of cells in G_1 and G_2 (Ordinate: numbers of cells in each DNA class). (Evans & Van't Hof 1974)

Three-dimensional reconstructions from serial sections of wheat nuclei photographed in the electron microscope have shown that in the interphase nucleus the chromosomes are not just mixed up like a tangle of string but are in a specific spatial arrangement, held in place by attachments to the nuclear membrane. This raises the possibility, as yet unexplored, that this arrangement could change during differentiation and that different cell types or organs could have their genome in a different spatial arrangement, so bringing different combinations of genes next to each other and perhaps promoting new gene interactions. But this is only speculation.

8.4   SUMMARY

(1)   Cells produced by the apical meristems cease division as they leave the meristem. They vacuolate, enlarge, differentiate, and stop growing. This process is cell maturation. Why cells stop dividing is not clear but it might be because they are displaced away from some source of division-promoting growth regulator in the meristem.

(2)   Cell enlargement and maturation is accompanied by synthesis of proteins, nucleic acids, and other cytoplasmic constituents, as well as synthesis of wall components. It has been analysed most in the root, where increasing basipetal distance of cells from the apical meristem represents a developmental sequence in time. Although there are changes in the protein complement and in the relative activities of enzymes during cell enlargement, these seem to be quantitative rather than qualitative and may be at least partly because of the changing proportions of various tissues as the root or stem develops. There is no firm evidence for the synthesis of new proteins. Cell enlargement may involve programmed changes in gene expression or it may be the result of the synthesis of different proteins at different rates set at the beginning of cell maturation. The cells seem autonomous for maturation since it continues in isolated root sections.

(3)   The mechanism of cell elongation has been studied mostly in those excised tissues where it can be promoted by auxin. Auxin acts in the first hour or so by increasing proton efflux from the cell into its wall. This causes wall loosening. Auxin also acts by altering gene expression, as shown by the change in the cells' protein complement. Auxin is the only growth substance that consistently increases wall extensibility. The cells that respond to auxin are the epidermal cells, the growth of which controls the rate of extension of the stem. The inner tissues have the potential to expand but are constrained by the epidermis.

(4)   Cessation of cell elongation is associated with increase of the wall protein extensin, which cross-links wall polysaccharides. Increased cross-linkage of the wall polysaccharides to each other may also contribute to decreasing wall extensibility and so too may an increase in auxin oxidases. As cells mature some may become polyploid but most stop growth in the G_1 or G_2 phases of the cell cycle, the proportions of G_1 and G_2 cells being characteristic of the species.

(5)   Plant cell differentiation most noticeably involves changes in cell wall structure and the amount and nature of the cell contents. The deposition of patterned wall thickening is particularly characteristic of xylem elements and has been studied in *Zinnia* cell suspensions. The positions of the bands of cellulose thickening are first marked out by the microtubules forming an underlying pre-pattern. Convoluted wall thickening in transfer cells can be induced by sugar solutions.

(6)   Cell wall breakdown is characteristic of phloem sieve plates and xylem vessels. In the latter, the end walls, which lyse, may have a different chemical composition from the side walls; this may facilitate selective lytic enzyme action. Lysis of cell contents is partial in the phloem sieve tubes but is complete in xylem elements and in sclerenchymatous fibres. Symplastic communication between cells may also decrease as they differentiate, and plasmodesmata are lost or obliterated by wall growth.

(7)   The changes in cell wall structure are accompanied by changes in wall composition. As the wall thickens, especially in secondary vascular tissues, pectin synthesis stops and the synthesis of hemicelluloses increases. The epimerase enzymes mediating the transformations of the UDP-sugars change in activity but are always active to some extent and do not seem to be potential control points. But the activities of the sugar synthase enzymes correlate well with the rates of synthesis of the wall polysaccharides and these synthase enzymes are, therefore, thought to be the main controlling agents for wall composition. The synthases may themselves be regulated at the levels both of gene transcription and translation (protein synthesis).

FURTHER READING

Barlow, P. W. 1978. Endopolyploidy: towards an understanding of its biological significance. *Acta Biotheoretica* **27**, 1–18.

Barlow, P. W. 1982. Root development. In *The molecular biology of plant development*, H. Smith & D. Grierson (eds), 185–222. Oxford: Blackwell Scientific Publications.

Barnett, J. R. (ed.) 1981. *Xylem cell development*. Tunbridge Wells: Castle House Publications.

Cutter, E. G. 1978. *Plant anatomy. Part I. Cells and tissues*, 2nd edn. London: Edward Arnold. (Cell types and their differentiation)

Fry, S. C. 1988. *The growing plant cell wall: chemical and metabolic analysis*. New York: Longman.

Green, P. B. 1976. Growth and cell pattern formation on an axis: critique of concepts, terminology, and modes of study. *Botanical Gazette* **137**, 187–202. (Interpretation of what actually happens during cell development)

Heyes, J. K. & R. Brown 1965. Cytochemical changes in cell growth and differentiation in plants. *Encyclopedia of Plant Physiology*, Vol. XV/I, W. Ruhland (ed.), 189–212. Berlin: Springer.

Northcote, D. H. 1985. Control of cell wall formation during growth. In *Biochemistry of plant cell walls*, C. T. Brett & J. R. Hillman (eds). SEB Seminar Series **28**, 177–97. (Biochemistry of callus and of secondary growth)

Robards, A. W. (ed.) 1974. *Dynamic aspects of plant ultrastructure*. London: McGraw-Hill. (Ultrastructural aspects of cell differentiation)

Silk, W. K. 1984. Quantitative descriptions of development. *Annual Review of Plant Physiology* **35**, 479–518. (Mathematical analysis of growth)

Taiz, L. 1984. Plant cell expansion: regulation of cell wall mechanical properties. *Annual Review of Plant Physiology* **35**, 585–657.

NOTES

1   Barlow (1976), Pilet & Barlow (1987). (Control of cell growth)

2   Allan & Trewavas (1986), Fowler & ApRees (1970), Navarette & Bernabeu (1978). (Changes in protein complement)

3   Barlow (1987) (Packet analysis); Erickson & Goddard (1951) (Kinetic analysis); Hejnowicz & Brodski (1960) (Cell lengths as time markers).

4   Cleland (1986) (Auxin and wall extensibility); Kutschera *et al.* (1987) (Epidermis); Masuda & Yamamoto (1985) (Changes in wall chemistry during cell extension); Rubery (1987) (Auxin transport).

5   Barlow (1984) (Root cap); Francis (1978) (Initiation of laterals after root decapitation); Heyes & Vaughan (1967) (Isolated root segments); Lyndon (1979) (Cell maturation and time).

6   Stafstrom & Staehelin (1988) (Extensin); Vaughan (1973) (α, α'-Dipyridyl).

7   Gunning & Pate (1974), Henry & Steer (1980) (Transfer cells); Roberts *et al.* (1985) (Microtubules); Smart & Amrhein (1985) (PAL and xylogenesis).

8   Benayoun *et al.* (1981). (Xylem wall breakdown)

9   Lyndon & Robertson (1976). (Ultrastructural changes)

10 Gunning (1978). (Plasmodesmata)

11 Jensen & Ashton (1960) (Onion roots); Northcote (1963) (Secondary xylem).

12 Gahan & Bellani (1984). (Cytochemical markers)

13 Barlow (1985) (DNA endoreduplication); Bennett (1984) (Wheat nuclei); Evans & Van't Hof (1974) (Proportions of cells in G_1 and G_2).

CHAPTER NINE

Genes and development

9.1 GENIC REGULATION OF DEVELOPMENT[1]

Clearly, genes can regulate development because plant form, such as leaf shape and flower structure, is inherited and differs between species. The necessity of certain genes for development is also shown by mutants such as the series of embryo-lethal mutants in *Arabidopsis*, which arrest embryo development at various stages. However, the mechanism of action of these and similar genes is not known. We do not know whether they code directly for enzymes that are part of the developmental programme or indirectly for enzymes involved in the synthesis of regulator substances, i.e. growth substances, controlling metabolites, or regulator proteins.

There must be close control of the changes in enzyme activities and in protein complements in different tissues and cells so that development is orderly and so that programmed differentiation pathways are followed. This means that in some way there is differential expression of the genes in different cells. For example, expression of genes for flower development is suppressed in the vegetative plant, and genes for seed proteins are expressed only in the seed. Gene expression may be regulated at various levels within the cell – transcription, post-transcriptional modification of the mRNAs, mRNA breakdown, translation (protein synthesis), protein turnover and breakdown, enzyme compartmentation within organelles, and enzyme activation.

9.2 GENE EXPRESSION DURING DEVELOPMENT[2]

DNA–RNA hybridization experiments show that there are approximately 50–100 000 genes expressed in a higher plant. Higher plants with the smallest amounts of DNA are those such as *Arabidopsis*, which has about 0.1 pg of DNA per haploid genome, and is a small annual with all the organs and organization found in larger plants, except for secondary thickening. The *Arabidopsis* genome consists of 70 000 kb (kilobases), so there must be approximately 2 kb per gene on average. Some of this DNA will be regulatory DNA, and introns that are spliced out before the gene transcripts become activated. Most higher plants have much more DNA than this in their genomes (the extreme is *Fritillaria*, with >100 pg) but probably express about the same number of genes as *Arabidopsis* does.

Experiments with tobacco have shown that about 25 000 genes are expressed in each organ type. About 8000 of these are common to all the organ types tested (i.e. leaves, stems, roots, petals, anthers, and **ovaries**) and about 6–10 000 genes are specific to each organ type. The rest of the genes are expressed in more than one organ but not in all. In most organs, some of the RNA transcripts detectable in the nucleus are not detectable in the organs, implying post-transcriptional regulation of gene expression. But in the ovaries, all the ovary-specific mRNAs were detectable in both the nuclear transcripts and in the cytoplasm, indicating that the main control of gene expression in the ovary tissues is at the transcriptional level. The contrary case is that of the stem, in which stem-specific mRNAs are not detectable in the nuclear transcripts, implying that the main control of gene expression in the stem is post-transcriptional.

Nuclear run-off experiments have shown that genes for seed and non-seed proteins may be transcribed at similar rates, although the proteins they code for may differ 1000-fold in abundance in the cell, again indicating post-transcriptional control of gene expression. In peas, the mRNAs for the seed proteins increase in amount just before and during the accumulation of the seed proteins themselves and so these two sets of events are correlated, consistent with gene expression being regulated at the transcriptional level. For one protein, the mRNA level remains high when the synthesis of the protein has stopped and so in this case there must also be some post-transcriptional control. Genes may, therefore, be considered as not only switched on or off but also up-regulated and down-regulated. Soybean seed proteins are not present in mature plant organs other than the seed, or at least are not detectable as being present at a level greater than one molecule per 100 cells. They are essentially switched off in all but the seed cells.

Changes in gene expression during embryogenesis have been monitored by seeing how the population of mRNA changes during development. mRNA is extracted from embryos at different stages of development and is used as the template for protein synthesis in an *in vitro* system (usually extracted from wheat germ). Separation of the proteins by two-dimensional polyacrylamide gel electrophoresis (PAGE) allows the composition of the protein complements and, therefore, the mRNA complements to be compared. In cotton embryogenesis, groups of mRNAs change in abundance as development proceeds and appear to

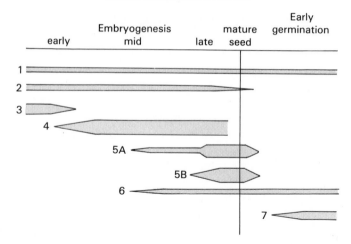

Figure 9.1 Changes in abundance of groups of mRNAs during cotton embryogenesis. Group 1 represents genes expressed all the time. Group 4 represents the genes for the seed proteins. Functions of the proteins coded for by other groups of mRNAs are not known. (Dure 1985)

be stage-specific (Fig. 9.1). The functions of the proteins produced by most RNAs is not known but we do know the function of some of the constitutive genes (those expressed all the time), such as the genes for tubulin and calmodulin, and those that code for the seed proteins.

9.2.1 Gene promoters and control regions

Upstream in the DNA of each gene for organ-specific proteins there are promoter sequences that seem to be necessary for the genes to be expressed. This has been shown by transfer of genes by means of plasmid vectors from a donor plant to a receptor plant, which is therefore a transgenic plant. Important regulatory sequences seem to be contained in the DNA −500 bp upstream from the start of the gene itself. If this upstream DNA is first deleted and then the gene is inserted into a transgenic plant, the gene is no longer expressed. This has been shown for seed proteins, which are expressed only in the seed or else at very low levels in other organs. In the soybean seed protein DNAs, there are regulatory and control sequences in the region −1000 bp upstream from the gene in each case (Fig. 9.2). The pea seed protein DNAs have a 'legumin box' in the region −80 to −120 bp upstream (the box itself is a 28 bp conserved sequence) and a 'vicilin box' of a conserved sequence of 42 bp approximately −120 to −170 bp upstream of the genes themselves. These sequences are identified as possible regulatory sequences because of their similarity to sequences which have been shown to be essential in other seed protein genes.

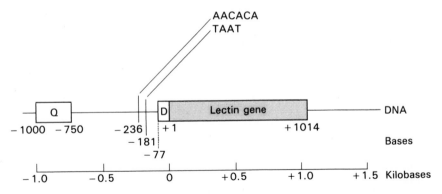

Figure 9.2 Control sequences in the DNA upstream of the soybean seed lectin gene. Developmental (D) and quantitative (Q) regions are important for the expression of the lectin gene in the seed. Q is necessary to raise lectin mRNA levels by 1000-fold during seed development and also in the mature plant root. The AACACA consensus sequence is another possible regulatory sequence that is common to many other seed protein genes. The TAAT region interacts with an embryo DNA-binding protein. (After Goldberg 1989)

When soybean seed protein gene DNA was transferred to transgenic tobacco, the soybean seed proteins were detected in the tobacco plant but only in its seeds. Furthermore, when soybean seed protein DNA was transferred together with DNA coding for three non-seed proteins, the latter were expressed, as expected, not only in the seed in the tobacco but in other organs too, whereas the soybean seed protein was expressed only in the tobacco seed. This experiment indicates that the regulatory sequences work only for their own genes and not for other genes transferred with them. DNA for soybean seed proteins has also been transferred to transgenic *Petunia*. When the region of the DNA transcript −69 to −159 bp upstream of the gene was deleted before transfer, then the gene was not expressed at all in the *Petunia*, but when this stretch of DNA was present, the soybean protein gene was expressed in the *Petunia* seeds. These experiments show that:

(1) the same regulatory sequence is recognized in different species, i.e. there is probably a common conserved regulatory sequence and this is necessary for expression of the gene;
(2) the organ-specific expression is conserved between species and, therefore, the regulatory region is regulated according to the type of organ that the gene finds itself in.

The temporal aspect of expression can also be conserved between species. This was shown by the expression of *Phaseolus* seed protein in transgenic tobacco 15 days after pollination, as it would have been originally in the bean even though the native tobacco seed proteins began to be synthesized after nine days. Thus, the bean protein gene conserved not only its organ-specific expression in the tobacco but also the time of its expression relative to pollination, even though this meant that it was out of step with the synthesis of the tobacco seed proteins in the host tobacco cells. Attempts to get expression of cereal seed proteins in transgenic dicotyledonous plants have not yet succeeded. This may be because the expression of the genes is tissue-specific and, at least in the legumes in which the experiments have been tried, the seed proteins accumulate in the cotyledons rather than in the endosperm, which is a major component of cereal grains but is only a short-lived tissue in most dicotyledons.

The activation of genes by light has also given evidence for an upstream regulatory sequence, which is recognized universally in plants. Genes that are light-activated retain this requirement when transferred from one species to another. This has been shown for a wheat *cab* (chlorophyll *a/b*) gene transferred into tobacco where it retained its requirement for light activation and also its tissue specificity, showing also that the same regulator elements are recognized in monocotyledons and dicotyledons. Similarly, the gene coding for the small subunit of Rubisco (ribulose bisphosphate carboxylase) protein from pea is also expressed in the light in transgenic *Petunia*. Expression of genes can be cell-specific and tissue-specific as well as organ-specific. But how are the genes regulated so precisely and even different genes in the same gene family switched on at different times? Somewhere in the genome there are presumably sets of regulator genes that produce regulatory molecules in a programmed sequence so that the promoters of individual genes are activated at the appropriate times during development. We can envisage an upstream regulatory region being an essential component of most genes and specifying organ, tissue, and cell specificity, as well as time specificity.

9.2.2 Gene regulation[3]

As well as being expressed or not, i.e. switched on or off, a gene may be expressed to a degree that may vary over several orders of magnitude. Up- and down-regulation seems especially to be the result of the action of growth substances or environmental conditions. For instance, when sulphur is in limiting supply, seed proteins rich in sulphur amino acids tend to be synthesized to a lesser extent than sulphur-poor proteins. The lack of sulphur results in a reduction in the amounts of mRNAs coding for the sulphur-rich proteins and is because of increased RNA turnover rather than decreased transcription. This post-transcriptional regulation could be occurring at the level of translation (protein synthesis), where mRNAs not being used because of unavailability of sulphur-containing

(a) **Multiple genes**

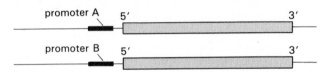

(b) **Multiple promoters**

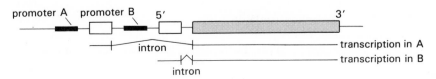

(c) **Multiple regulators**

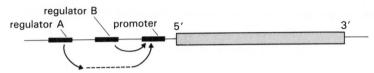

Figure 9.3 Possible alternative ways in which the same gene is expressed in different cell types or at different times during development. (a) Multiple identical genes but with different promoter sequences will be switched on by different regulators. (b) A gene may have several promoters in tandem, which are each activated by different regulators. (c) A gene may require different regulators (in *cis* arrangement as shown, or in *trans* arrangement). (Baulcombe *et al.* 1986)

amino acids, especially methionine, may be more liable to degradation than those actively involved in protein synthesis. So far, we do not have the data to know whether this is so.

When a gene is expressed at different levels at different times and in different tissues, its control can be under a more complicated control. The synthesis of α-amylase is controlled by a number of genes, some of which are expressed in various tissues throughout the plant but some of which are specifically switched on in the **aleurone** layer of the cereal grain. In this case, there are multiple genes coding for the same protein but with different promoter sequences, so that a different gene copy is switched on under different circumstances. There are also other α-amylase genes that are expressed in both the developing grain and in the aleurone and probably employ a different strategy, that of multiple regulator sequences arranged in tandem in the promoter sequence of the gene (Fig. 9.3). This means that the section of DNA producing the regulator that binds with the promoter region is located on the same stretch of DNA (in tandem) and is therefore in a *cis* arrange-

195

ment. It is believed that regulator sequences are usually *trans*, i.e. on another stretch of DNA in another chromosome or, if on the same chromosome, at least far removed from the gene in question.

The nature of the regulatory processes in higher plants is not yet known but is probably very like that in bacteria, where there are regulatory proteins coded by regulator genes elsewhere in the genome. However, it must be remembered that bacteria do not show the sort of developmental processes that are the hallmark of multicell-ular organisms. In some cases, the regulator genes of plants may be activated by light, via phytochrome or cryptochrome, or by growth substances, which may transduce the effects of environmental signals. In fruit ripening, five ripening-specific genes have been identified in tomato. These genes have been identified with the help of other mutant genes that prevent or delay ripening and appear to be mutations in the regulatory genes governing the expression of the ripening-specific genes. One of these apparently regulatory genes is the *rin* gene (*ripening inhibitor*). If this were simply a gene that affected the synthesis of ethylene required for ripening, then its effect should be overcome by supplying exogenous ethylene to the fruit, which induces ethylene production, or by wounding, but both of these treatments are without effect. The way in which the *rin* gene acts is, therefore, not yet known but it seems to switch off the ripening genes and may, therefore, code for a regulator protein. One of the ripening-specific genes is the gene for polygalacturonase (PG), an enzyme that breaks down pectic substances and so contributes to wall softening during ripening. The PG gene appears to be transcriptionally con-trolled. In the presence of the *rin* gene, the PG gene, although present, is not expressed. During fruit ripening, the protein complement changes quantitatively and possibly qualitatively, as it does in cell maturation (see Ch. 8, section 8.2) and in the shoot apex on flowering, implying a change in gene expression. How regulator genes concerned with controlling development are themselves regulated has not yet been discovered. Plant genomes contain transposable genetic elements, which are short pieces of DNA that can move around in the genome and become integrated temporarily into genes, thus prevent-ing or affecting the gene's expression. It is an interesting possibility that transposable elements are involved in the regulation of plant development, but so far there is no firm evidence to suggest that they are.

9.3   INDUCED CHANGES IN GENES GOVERNING DEVELOPMENT[4]

Some varieties of flax (*Linum*) grown for a single generation in phosphorus- or nitrogen-deficient conditions become much larger than normal, while the same varieties grown under optimal conditions are much smaller. These induced, but subsequently stable, differences in phenotype are paralleled by heritable differences in the genotype and so the different forms are called genotrophs. They can differ by up to 16% in DNA content per genome. This difference appears to be due to different copy numbers of the various repetitive sequences in the DNA, including the ribosomal genes. These genomic changes alter the development in a quantitative rather than a qualitative way, affecting degree of branching and plant size but not otherwise affecting the course of development.

9.4   GROWTH SUBSTANCES AND THE CONTROL OF GENE EXPRESSION

9.4.1   Gibberellin and amylase synthesis[5]

The control of the synthesis of α-amylase in the aleurone layer of cereal grains is probably the best-studied example of the regulation of gene expression by a growth substance, in this case gibberellic acid (GA_3). Gibberellic acid produced by the embryo on germination diffuses to the aleurone cells where it promotes the synthesis of α-amylase, which is then released into the endosperm where it degrades the starch to sugar. It is, therefore, a key process in the germination of cereal grains for the production of malt in the brewing industry. Gibberellic acid causes an increase in mRNA levels and, therefore, acts mainly by promoting the transcription of the α-amylase genes, although there may also be regulation at the translational (protein synthesis) level. The GA_3 action is indirect in that it does not interact itself with the DNA but apparently promotes the formation of regulator proteins.

There are two families of genes for α-amylase, each family being multigenic. Another enzyme that is synthesized in the aleurone in response to GA_3 is wheat carboxypeptidase. The DNA sequences coding for all these GA_3-activated genes have been examined for upstream sequences in common, which would then be candidates for regulator sequences, but no such consensus sequences have been found. The pathway by which GA_3 exerts its effect has yet to be discovered.

Other evidence from germinating bean seeds has shown that GA_3 can cause an earlier appearance of transcripts of the malate synthase gene and that abscisic acid (ABA) inhibits its transcription. The effects of gibberellins are often nullified by ABA, which, in general, has an inhibitory effect on plant processes. Abscisic acid prevents and retards germination and there is some evidence that ABA, like GA_3, regulates expression of some seed protein genes and not others. The precise mode of action of ABA is also not yet known.

9.4.2 Auxin regulation of gene expression

Auxin-regulated gene expression has been shown especially in soybean hypocotyls by the change in pattern of translatable mRNAs in an *in vitro* wheat germ translation system. Some mRNAs increase and an equal number decrease; altogether, about 10% of all hypocotyl proteins that are detectable are regulated by auxin. Nuclear run-off experiments have shown that auxin very rapidly (within 5–30 min) induces transcription of some RNAs. Much faster responses have also been found. By preparing cDNA clones to some of these RNAs that respond rapidly, and by using them as probes, it has been demonstrated directly that the levels of these mRNAs are regulated by the application of auxin to the tissues, although in some experiments this may be up to 12 h after the application of the auxin. The inhibition of auxin-induced transcription without inhibition of cell elongation during the first few hours is consistent with the early stages of elongation being a response to proton secretion, with only the later, sustained, stages of elongation depending on gene transcription and protein synthesis (see Ch. 8, Box 8.2).

In all the cases known so far, there is no evidence that the growth substances themselves act directly on transcription. They, and phytochrome and cryptochrome, presumably act by affecting the synthesis or release of proteins, which then act as regulators of the genes. Polypeptides shown to be specific gene regulators have not yet been isolated from plants.

It is important to note that the growth substances can have effects only when the tissues are sensitive to their action. This raises the whole problem of the nature of competence (see Ch. 10) and is illustrated by the quite different effects, or lack of them, that the same growth substance can have in different tissues and at different times during development. Gibberellins, for instance, act on aleurone cells but not on the neighbouring endosperm cells. Auxin acts on cell elongation in the stem epidermis but not in the pith (see Ch. 8, Box 8.2), and ethylene promotes fruit ripening only in mature fruit. The developmental programme is overriding – growth substances can only induce or repress those genes that are normally involved in the developmental programme at that stage of development in the organ in question. Whether or not growth substances are involved in the regulation of the developmental programme in the intact plant is not clear. Experiments suggesting that they could be are those showing that the induction of organogenesis in tissue cultures can be regulated by exogenously applied growth substances (see Ch. 10), but at the moment we do not know how, or how the effects of different growth substances might be integrated to produce the regular development from embryo to mature plant.

9.5 SUMMARY

(1) In a higher plant there are about 50–100 000 genes expressed and about 25 000 in each organ type. Some are organ-specific. The changes in gene expression during development, and especially during embryogenesis, have been monitored by following the changes in mRNA populations.

(2) Expression of some seed protein genes depends on the presence of regulatory sequences of DNA upstream from the genes. Isolated genes transferred to a host plant of a different species depend on their regulatory sequences for expression. The genes retain their organ-specific expression in the transgenic plants, showing that the base sequence of the regulatory region and the mechanism of organ-specific gene regulation are probably common to all higher plants.

(3) As well as being switched on or off, genes may be up- or down-regulated so that gene expression can be quantitative. The same gene may be expressed specifically in several organs or at several times during development. This can be achieved in several different ways: by multiple copies of a gene with different promoters on each copy; or by single copies with multiple promoters or regulators. The regulatory molecules that bind to the regulator sequences in the DNA are almost certainly polypeptides, but how the production and action of these in turn is regulated is not understood. Genes that prevent or delay fruit ripening are apparently regulatory genes.

(4) Gene expression can be controlled by growth substances, especially gibberellins, which can control the synthesis of α-amylase in the cereal grain aleurone cells, and auxin, which can alter gene expression during cell elongation. Abscisic acid may also be able to control gene expression, as well as acting as an inhibitor of development. The growth substances themselves do not seem to act directly as gene regulators but indirectly through their effects on the regulation systems.

FURTHER READING

Goldberg, R. B. 1989. Regulation of gene expression during plant embryogenesis. *Cell* **56**, 149–60.

Kuhlemeier, C., P. J. Green & N.-H. Chua 1987. Regulation of gene expression in higher plants. *Annual Review of Plant Physiology* **38**, 221–57.

Leaver, C. J., D. Boulter & R. B. Flavell (eds) 1986. Differential gene expression and plant development. *Philosophical Transactions of the Royal Society London, Series B* **314**, 343–500. (A valuable symposium covering most of this chapter)

Marx G. 1983. Genes for development. *Annual Review of Plant Physiology* **34**, 389–417.

Sheridan, W. F. 1988. Maize developmental genetics: genes of morphogenesis. *Annual Review of Genetics* **22**, 353–85.

NOTES

1 Meinke (1986). (Embryo-lethal mutants in *Arabidopsis*)
2 Dure (1985). (Embryogenesis)
3 Grierson (1986). (Ripening genes)
4 Walbot & Cullis (1985). (Flax genotrophs)
5 Rodriquez *et al.* (1987). (GA$_3$, ABA and gene expression)

PART V

Competence and determination

CHAPTER TEN

Competence and determination in differentiation

10.1 COMMITMENT

The development of different types of cells, tissues, and organs from common origins implies that:

(1) the cells are competent to differentiate;
(2) as development proceeds they lose the capacity to be easily transformed into other cell, tissue, and organ types.

The cells become committed, i.e. fixed or determined, in their development. Leaves do not normally form roots, or roots leaves, but if a leaf is cut from the plant and its petiole is placed in soil it may form roots at the end of the petiole. Roots of some species can be used for propagation because, when excised, they will form buds and eventually leaves. Such observations show that determination is either not absolute or it can be reversed in appropriate circumstances. They also show that competence to follow a different type of development either persists in the differentiated and determined tissues or is restored by the experimental treatment.

Competence to develop along a particular pathway means that the cells can go on to develop autonomously or can do so when given an appropriate signal. A state of competence can, therefore, be recognized when development proceeds as expected or when the cells are given the appropriate signal, which may be environmental, chemical, or some other experimental treatment. When competent cells begin to develop, the processes of development and differentiation become increasingly difficult to prevent or reverse. If the cells can be prevented from developing along a new pathway or can be diverted into a different developmental pathway, then the cells are not yet committed, i.e. they are not determined. Whether or not cells, tissues, and organs are

203

determined for a particular type of development can only be found by experiment. The fact that cells, when left to themselves, will develop autonomously in some particular way shows only that they are specified (in the terminology of animal embryology) and not that they are necessarily determined, which would imply that the commitment is irreversible.

10.1.1 Regional specification in embryos: fate maps and determination[1]

We have seen (see Ch. 1, section 1.3.4.1) that fate maps can be constructed showing which parts of the plant arise from which parts of the very young 16-celled embryo in *Capsella*. But this simply shows that cells in particular parts of the embryo will finish up in particular parts of the plant, e.g. the top part of the embryo forms the shoot and the lower part the root. It does not tell us whether there are differences, (except in position) between the cells at this stage, or when or whether different parts of the embryo have become determined for particular pathways of development.

In embryogenesis, the first sign of differentiation of cells in the proembryo (apart from the suspensor) is the formation of an axis marked by the formation of an axial procambial strand and by the protrusion of the cotyledons. The root and shoot apices also form at about this time. Whether the different parts of the globular proembryo are determined is not known. Experiments have not been done because of the difficulty of keeping isolated globular embryos alive and growing. There are several reasons for believing that the determination of the cells occurs only at the end of the globular stage. These are:

(1) Some gymnosperm embryos are, at first, free-nuclear and do not become cellular until a stage corresponding with the late globular stage.

(2) Early development of embryos is not always precisely the same in each embryo, but this does not seem to affect subsequent development. Early embryogenesis in cotton is like embryogenesis in cell cultures in having no regular pattern of division, although the embryos develop normally. A notable example is the Degeneriaceae: after the first transverse division of the zygote, development of both terminal and basal segments does not follow an exact pattern and can vary from plant to plant and be rather irregular, but this still leads to normal development. Another example in which the initial segmentation pattern varies is the gymnosperm *Sequoia* (in which there is no free-nuclear phase). This is like the ferns, where divisions are not regular after the quadrant stage and (as Bower noted 60 years

ago) 'there may be, and sometimes is, coincidence between the cleavages and the origins of parts, but the two processes do not stand in any obligatory relation one to the other'.

(3) In the *reduced* mutant of tomato, which lacks cotyledons, leaves, and a functional shoot apex, embryo development diverges from the normal at about the stage and time that the heart stage would form, i.e. at the end of the globular stage.

None of these reasons is really decisive, and in the absence of direct experiment (e.g. separation of the embryo into its parts to see whether each part can give a whole embryo) we cannot know at what stage determination occurs, although it seems likely to be about the time that an axis forms in the proembryo.

The *reduced* mutant of tomato is of interest in several other respects. Since it is elongated, it shows polarity and has an axis in the absence of cotyledons and shoot apex. It does not form cambium, although it can do so if a normal embryo with shoot apex and cotyledons is grafted onto it. It is, therefore, competent to form cambium but does not do so, showing that the stimulus for cambium formation is not present in the mutant. This implies that the shoot organs are determined before the vascular tissue and are required to form it. This is comparable to the situation in animals, in which the organ systems are mapped out first and tissue and cell types are determined later. In other tissues, T. Sachs has shown that a flux of auxin through the cells may induce the differentiation of vascular strands (see Ch. 7, section 7.6.2). If this is a necessary condition for vascular strand formation, then the *reduced* mutant embryos may have developed an axis but not a polar auxin flux. A polar transport system for auxin has been shown in embryos of unripe seeds of bean (*Phaseolus vulgaris*) and sycamore (*Acer pseudoplatanus*) in which procambium was already present. In the *reduced* mutants, the polar auxin transport system may not be present because the probable auxin sources, the shoot organs, do not form.

Since epidermis, cortex, and stele are distinguishable in the *reduced* tomato mutants, this shows that cells may become determined without first the formation of organs. In other plants, it has proved possible to prevent organ (leaf) formation by the application of phenylboric acid or by pricking the leaf primordia as they are about to form, but this has not prevented the development of vascular tissues. This is also consistent with the hypothesis that, in embryos, the tissues can become determined without organ formation first. The determination of tissues and organs may, therefore, normally go on in parallel but with the one not necessarily accompanied by the other. If this is so, then we might perhaps expect sometimes to find mutant embryos that can begin to form organs but not vascular tissues.

Is there any evidence of the order in which organ systems are determined in the plant embryo? Irradiation of embryos has shown that sensitivity (as shown by inhibited growth) varies inversely with the degree of development. In the early cereal proembryo, the **scutellum** and shoot are the most sensitive, but in the later embryo, these lose sensitivity and the root becomes sensitive to irradiation. This would be consistent with the cotyledons and shoot becoming determined before the root. In *Cuscuta*, the youngest cultured embryos proliferated callus throughout their length, but slightly older embryos formed callus only at the root pole and not at the shoot pole, again consistent with the shoot becoming determined before the root. In older embryos, the different parts were determined, since the radicle formed only radicle and the **plumule** formed only plumule.

10.2   IS THERE SUCH A THING AS AN UNDETERMINED CELL?

A completely undetermined cell would be one that could be diverted into any possible developmental pathway. In this sense, probably the only undetermined cells are the zygote and the cells of the early embryo, and callus cells or any cells that are embryogenic or can grow and produce a whole plant.

Once cells become part of an organized structure, it is doubtful whether they are any longer completely undetermined. The cells of a shoot or root of a moss can be traced back to the apical cell of that organ, but the apical cell is not entirely undetermined since it can give rise to either a root or a shoot but not both. Although meristematic cells are often regarded as undifferentiated, an isolated shoot meristem in culture produces a shoot and a root meristem in culture produces a root. The formation of roots on shoot meristems and shoots on root meristems involves first the formation of callus before the different type of organ is formed. The cells dedifferentiate and change their state of determination. Sometimes callus from the shoot is distinguishable from callus from the root. It is unusual for a shoot meristem to transform directly into a root meristem or vice versa, although some examples are known (see section 10.7). Parenchyma cells may be determined as such, but in the medullary rays they can re-differentiate into cambium (but see Ch. 7, section 7.7). Similarly, cortical or other cells can re-differentiate into cork cambium. The cells of the shoot may, therefore, be determined as shoot cells, with the parenchyma as shoot parenchyma, and yet still be capable of further determination as vascular tissues. This implies that determination in plants is a hierarchical process. This is also the case in animal embryos, in which Slack has suggested that 'cells which end up with the same histological type, but which are of different embryological provenance, are at least transiently "non-equivalent", i.e. exist in different states of determination'.

Determination is therefore not synonymous with differentiation. Differentiation, in the sense that it is used in animal development, is the formation of different cell types. In the animal, this is a more easily accepted concept because organs consist of specific cell types, e.g. liver cells or kidney cells, which are biochemically distinctive and which, in culture, perpetuate their kind. In the plant, organs are not distinguished by being made of specific cell types. Roots, stems, and leaves all possess the same kinds of cells (although there may be some specialized cells, such as mesophyll or stomata) but arranged in different ways and in organs having different overall shapes. The comparable situation in the animal is the formation of limbs or a head, but in the animal the differences in cell types between such organs are relatively great. In plants, the determination of cell types is, therefore, much more obviously superimposed on the determination of organ types and the two are more clearly seen as separate phenomena that coexist. Also, since plant vascular tissues are derived from procambium, and vascular strands can lack one of the major vascular tissue types (xylem) again the hierarchical nature of determination is perhaps more obvious in plants. Development may, therefore, be viewed as a sequence of decisions at which the cells become more and more restricted in their possibilities for development, with determination occurring at several steps. Once cells are determined and differentiated, this tends to be the end point of their development and it is unusual for one type of differentiated cell then to differentiate into another type.

The shoot meristem, though, does change from one type of functioning to another. The vegetative mode of growth is indeterminate, i.e. potentially endless, but when the transition to flowering occurs it changes to a determinate mode of growth, which then brings development for the flower meristem to an end. This highlights another difference between animals and plants. In plants, the germ line is not differentiated until the flower forms and so determination of germ cells is one of the later events in meristem development (see Ch. 2, Box 2.2). In animals, a cell may become determined relatively early in development but its nucleus may remain undetermined, as shown by nuclear transplantation experiments, perhaps suggesting that determination is primarily a property of the cytoplasm. In plants, this is less obvious because, under the appropriate conditions, probably all nucleated cells can be made to dedifferentiate and ultimately to become totipotent and produce a whole plant.

10.3   DETERMINATION IN CALLUS AND AT THE CELLULAR LEVEL

10.3.1   Callus as undetermined cells

Callus can be manipulated to grow in different ways, depending on what growth substances and other chemicals are supplied. Tobacco, especially, can be maintained as callus or can be made to differentiate roots or shoots, or both, according to the ratio of auxin : cytokinin in the medium. Some types of callus will not do this; they are apparently not competent to be affected. There is presumably an innate ability to be altered that is first required and can perhaps be produced by sufficient manipulation of the growing conditions. This requirement is called competence. Competent cells may then be affected in some way to alter their pathway of development or to initiate a new type of development. Once this has happened and the new mode of development is inevitable, then the cells are committed and have become determined. The process of commitment or determination may be recognized by the cells completing a new developmental pathway and not being diverted from it. Can we experimentally recognize and distinguish competence and determination? In some callus we can.

10.3.2   Competence and determination in callus[2]

Callus of *Medicago* continues to grow as callus on a medium containing 2, 4-D (an auxin) and kinetin (a cytokinin), but when transferred to growth-substance-free medium it can then initiate and form roots or shoots, depending on the ratio of auxin to cytokinin in the previous treatment. If the auxin : cytokinin ratio in the pretreatment was high, then shoots formed, and if low, then roots formed. (This is the opposite of the requirements in tobacco callus for shoot and root initiation.) Pretreatment for three or four days was optimum before transfer to the growth-substance-free medium; further pretreatment did not give any more organogenesis subsequently (Fig. 10.1). If the pieces of callus were too small, less than about 100 μm in diameter, they did not respond to these treatments. The maximal response was obtained with the largest pieces used, about 200–800 μm in diameter. However, the induction itself took place in a larger tissue mass, which was then split up and plated out to measure the organogenetic ability. So we know only that the final response required masses of not less than about 12 cells (which the smallest 105 μm pieces contained) and we cannot be sure whether induction might be possible in fewer cells.

208

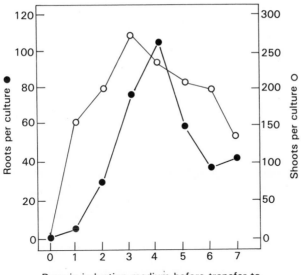

Days in induction medium before transfer to
regeneration medium

Figure 10.1 Induction of roots or shoots in *Medicago* callus. Shoots or roots formed only after transfer to regeneration medium (lacking auxins and cytokinins) following various periods beforehand on induction medium. The auxin : cytokinin ratio in shoot induction medium was low (5 µM 2, 4-D; 50 µM kinetin), and in the root induction medium was high (50 µM 2, 4-D; 5 µM kinetin). After 3–4 days, maximum competence had been acquired for subsequent organogenesis. (Walker *et al.* 1979)

These experiments show that:

(1) the callus was already competent to respond to the inductive effects of 2,4-D and kinetin;
(2) induction of organogenesis (determination) occurred during the pretreatments before the actual organ formation and required different conditions;
(3) competence to respond depended on the size of the cellular callus aggregates.

These experiments were done with already competent tissue, which could be induced to become determined to form roots or shoots. We do not know how determined the cells were because the possibility of reversal was not tested. Note also that the progress of determination, as the change in the state of the tissue, was observable only by transfer to different conditions. This was a true inductive change, i.e. one that required new conditions (a new signal) to be elicited.

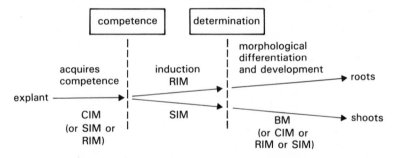

Figure 10.2 Induction of shoots and roots in *Convolvulus* leaf explants. Competence can be acquired on various media. Induction requires a specific inducer. Development then proceeds on basal medium (BM). SIM = shoot-inducing medium, RIM = root-inducing medium, and CIM = callus-inducing medium. (After Christianson & Warnick 1985)

10.3.3 Competence and determination in leaf explants[3]

Experiments with *Convolvulus* leaf explants have also allowed the experimental separation of the acquirement of competence for organogenesis from the actual induction of organs. Three different media were used: CIM (callus-inducing medium), SIM (shoot-inducing medium), and RIM (root-inducing medium). These media differed only in that the auxin : cytokinin ratio in CIM was high, in SIM was low, and RIM had high auxin but no cytokinin. Explants were placed on the appropriate inducing medium for different lengths of time and then transferred to basal medium for organogenesis to occur (Fig. 10.2).

Explants on SIM for <7 days regenerated negligible numbers of shoots when returned to basal medium, but >7 days exposure to SIM caused shoot regeneration, with 11 or more days on SIM giving maximal effect so that shoot induction required 7–11 days on SIM. Part of this time was concerned with acquiring competence, as shown by first placing the explants on CIM (which did not induce shoots) before placing them on SIM. This reduced the time required on SIM from 14 to 10 days (a different genotype was used for this experiment), and this 4 days was therefore assumed to be the time required for the explants to become competent and the further 10 days the time required for induction. The competent state was apparently stable, as shown by the fact that even 14 days on CIM did not reduce the period required on SIM to <10 days. Different genotypes behaved differently. Some required most of the time to become competent and only a very short time for determination. For instance, genotype 23 had its requirement for SIM reduced to 3 days after preculture on CIM for 10 days (Table 10.1).

This system allows questions to be asked about the nature of the determination process. Is determination a single-step process giving either a root or a shoot, or does determination proceed by at least two

Table 10.1 Competence to form shoots in leaf explants of *Convolvulus* genotype 23. The time to acquire competence is shortened from 7 days to 3 days by pretreatment on callus-inducing medium (CIM) for 10 days before transfer to shoot-inducing medium (SIM). (Christianson & Warnick 1983)

	Mean number of shoots formed per explant					
			Days on SIM			
Days on CIM	0	3	5	7	10	12
0	0	0	0	0.7	4.8	4.9
3	0	0	0.7	3.3	8.0	7.9
5	0	0.6	5.6	4.8	5.9	7.1
7	0	0	3.4	5.1	7.4	7.6
10	0	1.4	1.8	3.8	9.4	1.0
14	0	1.9	2.8	3.8	6.2	5.0

steps, first for an organ primordium and then for root or shoot? If it is a two-step process, it should be possible to transfer explants from SIM to RIM part-way through the inductive process and get the formation of roots without increasing the total time spent on inducing medium. In fact (with genotype 30), transfers from SIM to RIM after the first 4 days or so (corresponding to the period of acquirement of competence) reduced the number of roots formed, and after 14 days on SIM only shoots were formed despite a further 7 days on RIM before transfer to SIM. These results are consistent with the competence to react being common to whatever subsequent organogenesis occurs but with the determination produced by SIM being strongly for shoot formation and not first for unspecified organ formation. Determination here seems to be a one-step process. Determination may be said to be strongly canalized, i.e. not easily diverted from the preferred pathway.

10.3.4 Callus and leaf explants

These experiments lead to the following conclusions:

(1) The acquirement of competence may require similar, but not necessarily the same, conditions as determination.
(2) The time to acquire competence and the time to become determined may each vary from a few to many days, according to genotype.
(3) Determination, once begun, may not easily be altered or reversed.
(4) The requirements for induction of determination are not the same as those for the actual processes of organogenesis.
(5) Determination for organogenesis in callus requires a group of a minimum number of cells.

It seems, therefore, that callus can be undetermined and can form roots, shoots, or indeed the whole plant. As we shall see later, there are also callus types that do in fact show determination in themselves. Such calli are those that are embryogenic and those that are not.

10.3.5   Callus as determined cells

Although determination is essentially a property of organized structures, the experiments with callus show that determination can exist in the unorganized, cellular state. Perhaps the best example of the determined state persisting in callus that does not form an organized structure is the crown gall, which is determined as callus and is so canalized that it cannot be made to grow in any other way. Another example is habituation, when cells become able to grow without added growth substances (Box 10.1). Experiments with other types of callus show that the determined state can last for many cell generations in the cellular state.

Box 10.1   Habituation – a paradigm for determination?[4]

Habituation is the process by which callus cultures requiring a particular growth substance become independent of an exogenous supply and are then able to grow without it. Habituation has been found only with respect to cytokinins and auxins. Other plant growth substances are usually not essential additions to the culture media, although they may sometimes be needed, e.g. to suppress organogenesis or to promote growth of embryos. Habituated cultures can either synthesize what they require or no longer have an essential requirement for the growth substance in question. Some pith cultures of tobacco, especially those from a tumour-forming hybrid, will grow in the absence of added growth substances and are, therefore, already autonomous (as crown gall is). Callus autonomous for auxin and cytokinin can also be obtained from a number of plants.

Habituation occurs spontaneously but very infrequently, like mutation, but habituation is about 100 times more frequent than mutation, i.e. about 10^{-3} per cell generation compared to about 10^{-5} for mutation. Habituation is, therefore, presumed to be epigenetic, or developmental, in nature. Furthermore, habituation can be reversed, by transfer to a medium containing relatively high concentrations of the growth substance in question, by regeneration of new organs in the culture, or by regeneration of a whole plant and passage through meiosis and the seed. Habituation can also be induced and reversed by transferring the cultures to media with different concentrations of the growth substance (Table 10.2). Other treatments inducing habituation, such as altering the temperature temporarily, can be interpreted as having the effect of altering the cellular concentrations of growth substances temporarily and so inducing a feedback loop or breaking it (Fig. 10.3).

Table 10.2 Reversal of habituation in tobacco callus. The degree of habituation, R, is the ratio of relative growth rate on medium without cytokinin (kinetin) to the relative growth rate on medium with cytokinin. The R values were obtained by growing callus with or without kinetin at each stage and in each treatment. R values <0.4 indicate non-habituated callus. Habituated callus transferred to 16° reverted to non-habituated. After culture in the presence of kinetin for 6 weeks at 25°C (+K) or maintained on high kinetin in the previous 16°C treatment, callus became once again habituated. (After Meins & Binns 1978).

Treatment for 10 transfers at 16°C	Degree of habituation (R)			
	Before transfer to 16°C	At time of transfer to 25°C	3 wks after 1st transfer at 25°C	3 wks after 2nd transfer at 25°C
low K	1.6	0.2	0[a]	−K 0[a] +K 1.01
high K	1.4	0.7	0.3	−K 1.70 +K 2.04

[a] On medium −K the tissue was dead.

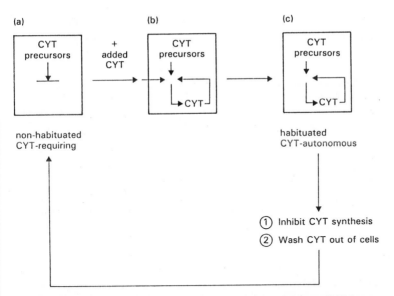

Figure 10.3 A model for habituation. The synthesis of cytokinin (CYT) is assumed to require the presence in the cell of CYT above a threshold amount. When non-habituated cells (a) are supplied with CYT then CYT synthesis is triggered and the CYT that is synthesized acts to promote its own synthesis (b). When CYT is no longer supplied (c), the cells can make their own CYT and are habituated. The feedback loop can be broken by reducing the CYT concentration in the cell either by inhibiting its synthesis or by washing it out. This scheme is consistent with many experiments. (After Meins & Binns 1978)

Is habituation relevant to normal growth and development?

The shift from one stable state to another, and back again, as happens in habituation and its reversal, can be illustrated by a model (Fig. 10.4) that is also appropriate for describing the transition between stable states that occurs when cells become determined. In that case, the reversal would be dedifferentiation. The possible involvement of habituation in the normal processes of development would be consistent with several observations:

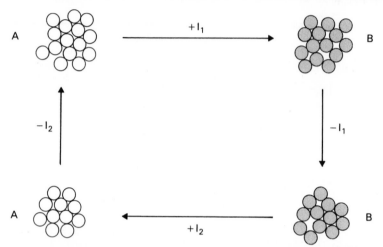

Figure 10.4 A model of the transition between two stable states A and B. The inducer I_1 causes the transition from A to B. State B remains stable when the inducer is removed. Inducer I_2 causes the transition B to A, A being stable when the inducer is removed. States A and B could be, for example, non-habituated and habituated callus, callus and a whole plant derived from it, or vegetative and flowering states. (Meins & Binns 1979)

(1) Tissues from different parts of tobacco plants show different degrees of habituation when cultured (Table 10.3).

(2) The degree of habituation increases up the plant and is greatest towards the shoot tip. This is correlated with the distribution of meristematic cells and the ability of the tissues to form buds in culture (Fig. 10.5).

(3) The shoot apex behaves like a tissue that can become habituated for auxin synthesis, since young leaves are sources of auxin and they arise on the apical dome, which cannot synthesize its own auxin (see Ch. 2, section 2.2.2.2). The field theory of phyllotaxis (see Ch. 3, section 3.2.2) is consistent with centres of inhibition of the formation of new primordia arising at minima in the supposed inhibitor field at the apex. If this theory were correct, this would mean the induction of the formation of a substance at minimal concentrations of that same substance.

(4) Like habituation, determination in development persists after the inducing stimulus has been removed.

Table 10.3 Habituation in callus from different tissues of tobacco plants. The degree of habituation, R, is the ratio of relative growth rate on medium without cytokinin (kinetin) to the relative growth rate on medium with cytokinin. R differs for callus of different origins. Pith callus, which is not habituated, becomes so after induction by cytokinin, whereas cortex and leaf callus are almost unaltered. (Meins & Lutz 1979)

Tissue	Degree of habituation (R) of callus	
	After one transfer on basal medium	After one transfer on cytokinin medium
stem pith	0.07	1.46
stem cortex	1.09	1.14
leaf	0.25	0.25

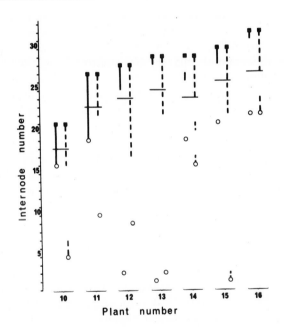

Figure 10.5 Correlation of habituation and flower formation in tobacco. Usually, only internodes at the tip of the plant formed habituated callus, and flower buds formed only on cultured epidermal explants derived from these same internodes. Dashed vertical lines indicate the extent of habituated callus and solid vertical lines the extent of *in vitro* flower bud formation. The top of the plant (■), the lowest branch of the inflorescence (—), and the lower limit of the region tested (o) are shown. Habituated callus is similarly formed from the upper part of juvenile plants, but flower buds are not. (Jackson & Lyndon 1988)

Habituation may be regarded as being the induction of the synthesis of a growth substance when it is at low concentrations within a tissue, and its repression when it is maintained at high concentrations. Whether or not habituation is indeed involved in normal development is not yet known, but it would be surprising if a process that occurs in cells of many species – and involves the two types of growth substance essential to maintain growth of plant cells in culture – were not part and parcel of the normal developmental metabolism. Even if it is not, it can provide us with a model that could be used to propose and test some hypotheses of metabolic control during development.

10.3.5.1 Determination in callus: embryogenesis[5]

The production of embryos in culture from somatic cells (sometimes called embryoids to distinguish them from zygotic embryos) occurs in some callus cultures but not in others. Some calli and cells are competent to form embryos and others are not. Embryogenesis occurs directly from competent cells, which have been called pre-embryogenic determined cells (PEDC). When the cells need to dedifferentiate and proliferate as callus before becoming embryogenic, then they are called induced embryogenic determined cells (IEDC). The hypothesis is that PEDC are already competent for embryogenesis and require only the appropriate conditions to become determined as embryo cells, whereas IEDC need to be made competent before they can become determined as embryo cells (Fig. 10.6).

Pre-embryogenic determined cells would include callus cells that form embryos directly, cells such as the epidermal cells of plantlets of *Ranunculus celeratus* grown from callus, and nucellar cells of citrus fruits, which form adventive embryos asexually. These latter examples also imply that determination as PEDC would coexist with determination as epidermal or nucellar cells.

A common observation is that in order for callus to become embryogenic it first needs exposure to 2,4-D and then needs transfer to a medium lacking 2,4-D for embryogenesis to occur. Such callus is assumed to be IEDC, the exposure to 2,4-D making it competent for embryogenesis on a different medium. This is directly comparable with the experiments on *Medicago* callus and *Convolvulus* leaf explants, in which competence for organogenesis was induced on one medium and organogenesis took place only on a different medium. The easiest way to obtain embryos is from young embryos themselves, for mature tissues usually require dedifferentiation and passage through callus before embryogenesis can be induced. The cells, therefore, seem to pass gradually from being PEDC to IEDC as they mature. Their degree of competence and determination for embryogenesis decreases. This would explain why embryos can form sometimes from a group of cells and sometimes from a single cell – it would depend on the degree to which neighbouring cells differed in the competence for embryogenesis (Fig. 10.6).

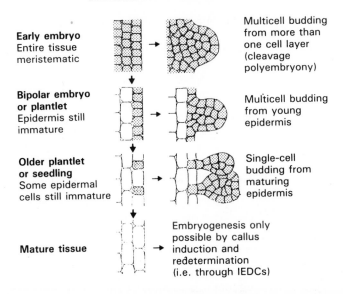

Early embryo
Entire tissue
meristematic

Multicell budding
from more than
one cell layer
(cleavage
polyembryony)

**Bipolar embryo
or plantlet**
Epidermis still
immature

Multicell budding
from young
epidermis

**Older plantlet
or seedling**
Some epidermal
cells still immature

Single-cell
budding from
maturing
epidermis

Mature tissue

Embryogenesis only
possible by callus
induction and
redetermination
(i.e. through IEDCs)

Figure 10.6 Differences in the embryogenic ability of cells. Pre-embryonic determined cells (PEDC) can form embryos but probably lose this ability as they mature and then require induction via callus to become induced embryonic determined cells (IEDC). PEDC (shaded) probably become reduced in number and distribution as the embryo cells mature. (Williams & Maheshwaran 1986)

The induction of embryogenesis in carrot cell cultures is linked to the appearance of two new proteins specific for embryos and the loss of two proteins specific for callus. However, this is probably not a change from IEDC to PEDC but a change in the types of cells formed since the callus was obviously embryogenic, as all cells could be made to form embryos. These protein changes are probably the first stages of differentiation of the embryo cells. Non-embryogenic lines of callus did not contain these proteins and could not be manipulated to produce them. It remains unclear whether these proteins are actually required for competence and determination or whether they are simply markers of these processes. The two types of callus, embryogenic and non-embryogenic, again show that different states of determination, which can be regarded as a type of differentiation, can exist at the cellular level without overt differences between the cells.

10.3.5.2 Determination in callus: juvenility[6]

Juvenility in ivy (*Hedera helix*) can be recognized by a whole range of characteristics (Table 10.4). When callus cultures are made from juvenile or mature ivy, the callus from one can be distinguished from the other. Callus from juvenile plants grows faster and has larger cells than callus from mature ivy. In addition, when plants are regenerated from callus

Table 10.4 Characteristics of juvenile and adult ivy. (Modified from Hackett *et al.* 1987)

Juvenile	Adult
climbing or spreading	upright or horizontal
non-flowering	flowering
adventitious roots on stem	no adventitious roots
cuttings root easily	cuttings root poorly
anthocyanin present	very little anthocyanin
lobed leaves	entire leaves
1/2 phyllotaxis	2/5 phyllotaxis
plastochron 4.2 days	plastochron 3.2 days
shoot apex 140 μm wide	shoot apex 200 μm wide
large cells in apex and callus	small cells in apex and callus
faster callus growth rate	slower callus growth rate
faster rate of internode growth	slower rate of internode growth

the juvenile callus regenerates juvenile plants and the mature callus regenerates mature plants. Also, under appropriate conditions, callus from mature stems will regenerate embryos whereas callus from juvenile plants will regenerate shoots. Although different investigators have used different culture conditions and have obtained different results, the important point is that the juvenile and mature states are propagated at the cellular level and the cells, therefore, show different states of determination.

So far, it has not proved possible to convert juvenile into mature tissues by experimental treatment. On the plant, this occurs spontaneously after it has been growing for some years and is climbing and has plenty of light. Treatment with abscisic acid (ABA) stabilizes (but does not induce) the mature form. The effect of ABA is perhaps not too surprising since it generally antagonizes the effects of gibberellin (GA), which can cause reversion of mature tissues to juvenile. When adult ivy plants are made to revert to juvenile by GA, the state of determination at the cellular level can be tested by taking samples of the stem at intervals as the shoots revert, making them form callus, and then measuring the growth characteristics of the callus. In such experiments, the first characteristic to revert was the callus growth rate. Then, the juvenile characteristic of increased rootlet formation was gradually acquired by the cells over an 11-week period (Fig. 10.7). This shows that:

(1) GA, when given once, can alter the state of determination of the cells;
(2) the change in determination takes several weeks and therefore is gradual or continuously quantitative and not all-or-nothing;

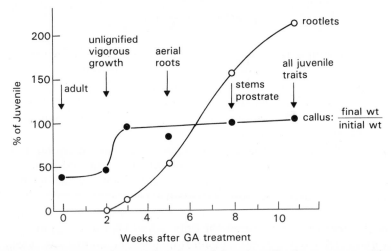

Figure 10.7 Reversion of ivy (*Hedera*) from adult to juvenile after treatment with gibberellin. Callus was derived at weekly intervals from adult plants sprayed with potassium gibberellate solution at time zero. The growth rate of the callus and the number of rootlets that it formed over a 4-week culture period were compared with callus obtained and cultured for the same time from juvenile plants. The changes at the cellular level were among the first observed. (After Miller & Goodin 1976)

(3) the different states of determination of juvenile and mature are properties of cells and do not require organized tissues for their propagation.

Similar experiments have been done with tobacco (*Nicotiana*). Callus derived from tissues of juvenile plants not yet able to flower regenerated only vegetative buds in culture, whatever the culture conditions. Callus from older plants, which were mature and flowering, could be made to regenerate vegetative buds or, if glucose was added, flowering buds. Thus, two types of callus had been obtained, one type competent only for vegetative growth and the other competent for flower formation as well if the appropriate signal, glucose in the medium, was given. Again, the transmission of the state of determination for vegetative or flowering growth was at the level of the cell. Other experiments with tobacco cultures have shown that the competence to form flowers or not is retained in thin cell layers, consisting of epidermis and a few cortical cells. The determination as juvenile or mature, and vegetative or flowering, is therefore a property of all the cells of the shoot and not some specialized cells. These cellular states of competence and determination may differ in different parts of the plant (Fig. 10.5).

10.4 COMPETENCE AND DETERMINATION IN XYLEM CELL DIFFERENTIATION[7]

A striking example of competence for xylem formation is the case of the fern prothalli which do not normally contain any xylem but in which xylem can be induced by the application of auxin or sucrose (see Ch. 7, section 7.4). This means the cells are competent to form xylem but normally the signal to elicit xylem differentiation is not present.

The process of xylem determination has been studied in lettuce (*Lactuca*) pith explants. Both IAA (an auxin) and zeatin (a cytokinin) were required for xylem induction, and the explants required exposure to inductive medium for five days for maximum xylem formation. However, by transferring the explants from inductive medium to non-inductive medium low in either auxin or cytokinin, it was shown (Fig. 10.8) that zeatin was required for only three days whereas auxin was required for five days. Serial transfer from a low auxin to a low cytokinin medium (without any exposure to a fully inductive medium) for five days on each gave only small numbers of xylem cells. The conclusion was that auxin and cytokinin were required simultaneously for the beginning of induction and that the requirement persisted for auxin but not for cytokinin. Since there was cell division on all media, the suggestion is that cell division, which may be required for the acquirement of competence, occurs first then determination occurs in 3–5 days followed rapidly in 1–2 days by xylem differentiation.

The time for xylem determination has also been measured during the induction of vascular strands with auxin (see Ch. 7, section 7.6.2), which had to be present for 2–3 days in order for xylem to become determined, as shown by its subsequent differentiation.

10.5 COMPETENCE AND DETERMINATION IN BUD INITIATION[8]

The protonema of a moss (*Funaria*) forms buds that grow into gametophores bearing the sexual organs. Under experimental conditions, untreated protonemata do not make buds for 72 h but bud formation can be promoted earlier if cytokinin is present. Only certain cells of the protonema respond to cytokinin, the apical three cells and the older basal cells do not, so only cells that are neither very young nor very old are competent to respond. Also, cytokinin has to be present for >12 h to have any effect. The number of buds is reduced if the cytokinin is removed between 12 and 72 h, the reduction being less the later the cytokinin is withdrawn (Fig. 10.9). Only after 72 h, when all buds have formed, does cytokinin removal no longer have any effect. Cytokinin

220

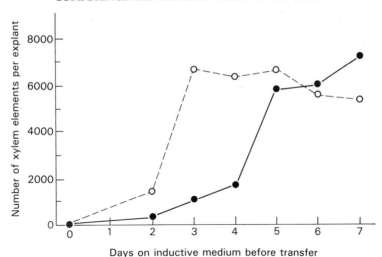

Figure 10.8 Different requirements for auxin and cytokinin for induction of xylem tracheary cells in lettuce (*Lactuca*) pith explants. The explants were first placed on induction medium (containing 10 mg l^{-1} IAA and 1 mg l^{-1} zeatin) for up to 7 days and at various times were transferred to either low auxin medium (-●-) or low cytokinin medium (-o-). They were sampled for xylem cell counts on Day 7. A shorter exposure to induction by cytokinin than by auxin was required for xylem formation. (Tucker *et al.* 1986)

must, therefore, be present continuously for determination to be completed. The determination process takes place over the period of 24–72 h and, therefore, takes an average of about two days. The need for cytokinin in this system may be compared with the need for the absence of auxin in embryogenic determination in callus and the efficacy of gibberellin in ivy callus.

In the succulent plant *Graptopetalum*, the lower residual meristem at the base of the leaf does not normally produce buds. Buds can be induced there by cytokinin, which has to be present for two or more days for maximum effect, although the first bud leaf is formed only after 3.5 days. This again would be consistent with determination taking about two days. A notable feature of this system is that the bud forms only where the cytokinin is applied, and cytokinin applied over a larger area gives a larger bud. This shows that the responding cells were only those receiving the stimulus directly.

10.6   COMPETENCE AND DETERMINATION IN ANTHERIDIUM INITIATION IN FERN PROTHALLI[9]

In fern prothalli, antheridium initiation can be induced by antheridiogen, a substance chemically related to gibberellin, which stimulates the initiation of antheridia. The test for antheridiogen production is the

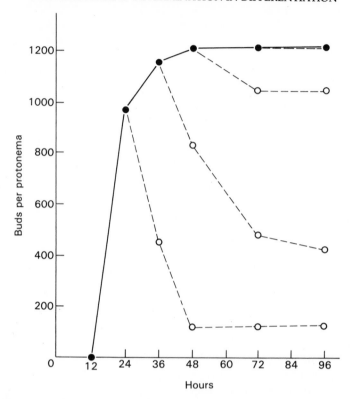

Figure 10.9 Induction of buds in the moss *Funaria* by cytokinin. Protonemata were grown on the cytokinin, benzyladenine (0.1 µM), for up to 72 h (-●-). At 24, 36, 48, and 72 h some protonemata were washed and transferred to basal medium (-o-). When protonemata were removed from the cytokinin at 24 h, most of the buds dedifferentiated. With later removal from the cytokinin, fewer buds dedifferentiated, until by 72 h removal from cytokinin had no effect because bud determination was complete. (Brandes & Kende 1968)

ability of the medium on which the fern gametophytes have been growing to stimulate antheridium initiation in thalli of *Onoclea sensibilis*, under experimental conditions where it has an absolute requirement for antheridiogen for antheridium initiation. At the filamentous stage, bracken (*Pteridium*) gametophytes are competent to initiate antheridia if supplied with antheridiogen but they do not produce it themselves (Table 10.5). At the next stage of development, the early heart stage, the bracken prothallus is assumed to begin to produce an inhibitor of antheridium induction and this seems to be made by the developing meristem, as the inhibition can be reversed by excision of the meristem. The thallus still does not produce antheridiogen itself, although at this stage it loses its sensitivity (competence to respond) to antheridiogen from other prothalli. This means that only young prothalli or, rather, small prothalli can initiate antheridia. At the late heart stage, the thalli

Table 10.5 Antheridium initiation in prothalli of bracken (*Pteridium*). (After Naf 1979)

Developmental stage	Sensitivity to antheridiogen	Formation of antheridiogen	Production of inhibitor of antheridium initiation
filamentous	yes	no	no
early heart	(yes)	no	yes
late heart	no	yes	no

stop producing antheridium inhibitor and start making antheridiogen. These older prothalli form archegonia.

The result is that in a population of prothalli, those that are faster growing do not get exposed to antheridiogen before they become insensitive, and so produce only archegonia. Slower growing prothalli are exposed to antheridiogen from the faster growing prothalli further on in their development and so form antheridia and later archegonia. The slowest growing prothalli are exposed to antheridiogen (from the faster growing prothalli) from early in their development and if their potential meristem cells become transformed early in their development into antheridia their further development is arrested and they become ameristic (lacking a meristem) and, consequently, show limited growth and do not produce archegonia. Apparently, no specific stimulus is necessary for archegonium initiation, which occurs once the prothalli have reached a later stage of development. The net result is that in a population of prothalli growing together, the fastest growers produce only archegonia, the average produce both archegonia and antheridia, and the slowest produce only antheridia. This promotes cross fertilization between neighbouring prothalli.

There are several points to note here. Production of antheridiogen and the hypothetical antheridium inhibitor, and competence to respond to antheridiogen (sensitivity) occur only at particular stages of development, showing that states of competence can change during development. Also, whether or not a particular prothallus makes antheridia or archegonia, or both, depends on its rate of growth in relation to that of its neighbouring prothalli. The interaction between prothalli could perhaps be a model for interactions between cells, in an organized tissue, that can grow at different rates and produce different stimuli at different stages of their differentiation. The occurrence of sex organs on a particular prothallus is the result of differentiation-dependent development analogous to that in the induction of vascular strands (see Ch. 7, section 7.6.2). There are two further points from these experiments. If antheridiogen is withdrawn from *Onoclea* prothalli before a late stage in antheridium formation, then the cells revert to having vegetative characteristics. Determination (at least in *Onoclea*), therefore, seems to occur gradually during the whole process of antheridium formation. The second point is that whatever the source of antheridiogen it does not affect the form of the antheridia, which is specific to each species. The antheridiogen, therefore, acts only as a trigger to development but does not prescribe the nature of the process that it triggers.

10.7   COMPETENCE AND DETERMINATION IN SHOOTS AND ROOTS[10]

In the induction of shoots and roots on *Convolvulus* leaf explants, the process seemed to be a single step to a particular type of primordium rather than first to an undefined primordium and then to shoot or root. Earlier experiments with *Convolvulus* root segments showed that the ratio of shoot : root buds formed depended on the culture conditions. Protracted auxin treatment promoted root initiation, which was suppressed by cytokinin. Shoot buds were formed mainly at the proximal ends of the root segments but also formed distally. Roots were only formed distally. It was not possible to tell whether roots and distal shoots could have a common origin or whether they were always distinct from the start of initiation. If auxin was withdrawn after a few days there was promotion of both shoot and root initiation. This suggests that in *Convolvulus*, as in the differentiation of xylem tracheary elements (see section 10.4), auxin is required both for the attainment of competence (in this case, to form primordia) as well as for determination (to form roots).

Roots or shoots may be initiated in tobacco thin cell layer explants but there is no clear evidence to suggest that there is a transformation of one type of primordium into another rather than a suppression of one type of primordium when conditions favour the other type.

There are some examples suggesting that young primordia can be undetermined initially and can then develop into either shoots or roots. Excised, cultured root tips of *Selaginella* were transformed directly into shoots when auxin was withheld or when the auxin transport inhibitor, TIBA, was present; and in *Nasturtium*, the axillary meristems, which normally grow as roots, can be transformed into leafy shoots by the application of cytokinin. In these cases, the acquirement of competence to form a primordium seems to precede determination as shoot or root primordium. The determination process can be influenced by auxin or cytokinin, with low auxin or high cytokinin tending to promote shoot determination.

10.8 COMPETENCE AND DETERMINATION IN LEAF DEVELOPMENT

10.8.1 Competence for leaf initiation[11]

The most striking example of the existence of competence for leaf initiation, although it is not normally expressed, is in those leafy liverworts that do not naturally possess ventral leaves. Two rows of ventral leaves can be made to form by growing the plants in the presence of hydroxyproline or α,α'-dipyridyl. These are antagonists of the synthesis of extensins – those cell wall proteins rich in hydroxyproline which increase in proportion in maturing cells and which, by cross-linking the wall components, could contribute to the cessation of cell growth (see Ch. 8, Box 8.3). When the synthesis of hydroxyproline in the proteins of the cell walls is inhibited, the walls probably remain able to grow for a longer period and the capacity for leaf formation is revealed. Although, in this case, it requires experimental treatment to make leaves form, presumably this could occur equally well if there had been an endogenous signal. A natural signal preventing leaf expression could be ethylene. In the presence of ethylene antagonists, the ventral leaves form apparently because the promoting action of ethylene on extensin synthesis is negated.

In the higher plants, the most widely held theories of leaf initiation assume that all points on the flanks of the apical dome are equally competent to initiate primordia and that the positions at which primordia arise are determined by the nature, age, and proximity of the existing primordia (see Ch. 3, section 3.2.2).

10.8.2 Determination of leaf primordia[12]

Many years ago, R. and M. Snow showed that the position of a leaf primordium on the apical surface could be altered by making incisions that isolated it from other primordia. Some primordia also developed as bracts rather than as leaves. Until half a plastochron before its emergence, the position of the primordium and its developmental fate could be changed, but after that, its position and development as a leaf became unalterable. Determination as leaf or bract, therefore, presumably occurred at the same time that the primordium position became fixed, i.e. half a plastochron before it became visible.

Determination of primordia in ferns is not complete until they are several plastochrons old. Fern leaf primordia can be cultured in isolation even when they have been excised just after initiation. The younger the primordium, the more likelihood there is that it will develop as a shoot (Table 10.6). By marking the apical cell with ink, Cutter showed that the

Table 10.6 Determination of leaf primordia of the fern *Osmunda*. Leaf primordia from the youngest (P_1) to that 10 plastochrons older (P_{10}) were excised and cultured on agar. The younger primordia can develop as shoots but as they grow older determination as leaves become more fixed. (Steeves 1966)

Primordium	Number of primordia (out of 20) developing as:		
	Leaves	Shoots	Doubtful or no growth
P_1	2	7	11
P_2	2	12	6
P_3	4	10	6
P_4	4	11	5
P_5	8	11	1
P_6	12	8	0
P_7	16	4	0
P_8	17	1	1
P_9	19	0	1
P_{10}	20	0	0

apical cell of the primordium in *Dryopteris* could become the apical cell of either a leaf or a shoot in those primordia that could develop into either. The young primordium is, therefore, not committed until later in its development. The primordium tended to develop as a shoot if it was excised before it was 2 plastochrons old. Older primordia could develop as either leaves or shoots, but in *Osmunda* the shoots seemed to develop from a new apical cell formed very near the primordium tip and not from the same apical cell as in *Dryopteris*.

Impatiens is a plant in which the terminal flower can be made to revert to vegetative growth, so that leaves are formed in the middle of the flower, by transfer from short to long days. If the plants are just on the point of being induced to flower and then transferred to long days, some of the primordia form organs that are intermediate between leaves and petals. These intermediate organs resemble small leaves but with patches of petal pigment, which shows that the primordium does not become determined as an entity, as one organ type or the other, but that different parts of the primordium can follow different developmental pathways. Because the smallest patch of petal pigment observed is about the same proportion of the modified leaf as one cell is of the original primordium, it appears that determination of organ type in this plant can go on locally at the level of individual cells or their immediate descendants. This is unusual. Normally, organs formed by the shoot apical meristem are determined as a whole and this may be because the shoot apex exercises a regulatory control that is absent in the case of *Impatiens*.

10.8.3   Determination of leaf form[13]

The process of determination of the different parts of the leaf has been shown to be a gradual one in pea leaves. Pea leaf primordia were cut or damaged at different stages in their early development (Fig. 10.10). If the surgery was done in the plastochron before the primordium became visible, the resulting leaf was normal, showing that complete regeneration of damaged parts had occurred. Thereafter, cuts affected the form of the whole leaf, the later cuts affecting the form of the leaflets. The leaf was, therefore, able to regulate its form only up to less than a plastochron after initiation, but regeneration of leaflets and more distal parts could occur a little later. These experiments show that the leaf was first determined as a leaf and that the whole process of leaf determination took about two days to complete, although determination of parts of the leaf may have begun later and so have been completed in a shorter time than for the leaf as a whole.

Some water plants, such as water crowfoot, are heterophyllous, the land and water forms of their leaves being distinctively different. The aerial leaves often have flat, lobed laminas but the submerged leaves are very finely dissected and filamentous. In such plants, the determination of leaf form takes place during most of the leaf's development. This is shown by the formation of leaves of intermediate type after transfer of plants from water to air, or *vice versa*. Only the very youngest primordium present at the time of transfer makes the full transition. The form of the water- and land-type leaves diverges only after the primordia are about 400 μm long and 3–4 plastochrons old in *Callitriche*. In this plant, the type of leaf developed can be controlled by abscisic acid, which induces land-type leaves on submerged plants, and by gibberellic acid (GA_3), which induces water-type leaves on emergent shoots.

It should be noted that although primordia such as leaves, sepals, and petals can be excised when quite small and will then grow, without added growth substances, into the expected type of organ (though often reduced in size and simpler in shape), this shows only that they are specified. In the absence of any experimentation that succeeds in altering their development, we do not know at what stage they become determined.

10.9   THE NATURE OF DETERMINATION

Determination can be recognized experimentally by the loss of the ability of cells, tissues, or organs to be diverted into a different developmental pathway. However, we do not know which cellular processes are important. Presumably they are involved with the stabilization of gene

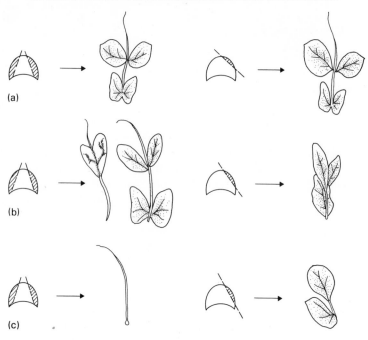

Figure 10.10 Determination in the pea leaf. Leaf primordia on the shoot apex of pea (*Pisum sativum*) seedlings were carefully damaged by having their sides or a portion at one side cut off. (a) When the primordia were operated on just as they were becoming visible as protuberances, they were able to regenerate the lost parts and formed normal leaves. (b) Primordia 30 μm long at the time of operation showed incomplete regeneration. (c) Primordia 70 μm long (1 plastochron old) showed no regeneration of parts removed. Determination of the primordia was therefore complete at about 1 plastochron (about 2 days) after initiation. (Sachs 1969)

expression. Whether determination involves the action of regulator genes is not known but it seems probable; how these genes themselves might be regulated is at the moment even more of a mystery.

10.10   SUMMARY

(1) Cells can be recognized as competent to develop if they do so autonomously or when they receive the appropriate signal. Cells may become specified for a particular type of development, according to their position in the plant, which implies that they will develop along a particular pathway unless they are diverted. The point at which cells or organs become determined can only be recognized by showing that they can no longer be experimentally diverted into an alternative developmental pathway. In the plant embryo, cells probably do not begin to become determined until the end of the globular stage and the beginning of organogenesis.

SUMMARY

(2) Except for the zygote, the cells of the globular embryo, and some undifferentiated callus, most or all cells are probably determined. Callus may be determined as embryogenic or root-forming or shoot-forming. Meristem cells are determined as root or shoot. Cells of young primordia at the shoot apex may be determined as leaf but not as specific cell types. Determination is, therefore, hierarchical in that cells may be determined for some characteristics and not for others.

(3) Experiments with callus and leaf explants have shown that competence for particular types of development precedes and can be distinguished experimentally from subsequent determination. The conditions for acquirement of competence and for determination can be different. Each of these processes seems, typically, to take a few days. In *Convolvulus* leaf explants, the competence of the cells seems to be common to whatever subsequent organogenesis occurs, but determination is strongly for roots or shoots and not first for an undefined primordium. Callus can often be made to be competent for embryogenesis by treatment with auxin (2,4-D), but for embryogenesis (determination) to occur it is necessary to transfer the callus to auxin-free medium. Embryogenic competence can be lost gradually as the callus ages.

(4) Natural changes in competence are shown by plants, such as ivy, with a juvenile phase. The young plants are determined as juvenile and the older plants as adult, each with many distinctive morphological and developmental features that can be transmitted through callus. Determination for juvenile or adult is, therefore, at the cellular level and does not depend upon organized supracellular structure. Each form, both juvenile and adult, is a stable state. Adult ivy can be made to become juvenile by treatment with gibberellin. Adult callus treated with gibberellin gradually assumes more and more juvenile features over a period of about 10 weeks, as shown by the characteristics of the plants that can be regenerated from it. So far, it has not proved possible to make juvenile tissues adult by experimental means.

(5) Experiments on the determination of roots and shoots, xylem, and buds in mosses and higher plants, have all shown that competence and determination can be under the control of plant growth substances and that different substances may be required for each step. Typically, the inducing substances need to be present throughout the whole period of acquirement of competence or determination, each of which often seems to take 1–3 days. Habituation of callus is a form of determination of particular interest because the changes in growth substance requirement can be induced simply by changes in the concentrations of the same growth substances themselves.

229

Whether changes in growth substance requirement can be induced in organized tissues by those same growth substances, and whether this could conceivably be involved in changes in competence and determination in normal development, is not known.

(6) In most plants, leaf primordia are probably determined as such at the time of their initiation, but in ferns, determination as leaf or shoot may be deferred for several plastochrons after initiation. Determination becomes more canalized with time, as shown by the greater difficulty with which it can be diverted. In pea leaves, determination is first as a leaf and then the various parts of the leaf gradually become determined in sequence during the next 1–2 days. Leaf shape, as shown by the water and land forms of leaves of heterophyllous aquatic plants, is not determined until the leaf primordia are several plastochrons old, and may be influenced by gibberellin or abscisic acid.

(7) In *Impatiens*, in which the terminal flower can be made to revert back to make leaves, organs intermediate between petals and leaves can be formed. They are leaf-like but with areas of petal pigment which can be restricted to small patches, implying that determination as leaf or petal occurs at a local level, perhaps in each cell, in this plant. In other plants, determination seems to be at the level of the organ as a whole and implies some integrating control by the shoot apex, which is lacking in *Impatiens*. The cellular and biochemical nature of the processes of competence and determination in any plant is quite unknown.

FURTHER READING

Meins F. & A. N. Binns 1979. Cell determination in plant development. *BioScience* **29**, 221–5.

Meins F. & H. Wenzler 1986. Stability of the determined state. *Symposia of the Society for Experimental Biology* **40**, 155–70.

Slack, J. M. W. 1983. *From egg to embryo*. Cambridge: Cambridge University Press. (Animal embryology, and concepts of development and determination).

Tran Thanh Van, K. M. 1981. Control of morphogenesis in *in vitro* cultures. *Annual Review of Plant Physiology* **32**, 291–311.

Wareing, P. F. 1987. Juvenility and cell determination. In *Manipulation of Flowering*, J. G. Atherton (ed.), 83–92. London: Butterworth.

NOTES

1 Caruso & Cutter (1970) ('Reduced' mutant); Fry & Wangermann (1976) (Auxin transport in *Phaseolus* and *Acer* embryos); Raghavan (1976) (Irradiation of embryos; *Cuscuta*); Wardlaw (1955) (Types of embryo development).

2   Walker *et al.* (1979). (*Medicago*)
3   Christianson & Warnick (1983, 1985). (*Convolvulus* leaf explants)
4   Meins & Binns 1978, Meins & Lutz (1979), Jackson & Lyndon (1988). (Habituation)
5   Sung & Okimoto 1983 (Embryo-specific proteins); Williams & Maheshwaran (1986) (Embryogenesis: PEDC and IEDC).
6   Chailakhyan *et al.* (1975), Hackett *et al.* (1987), Miller & Goodin (1976). (Juvenility in callus)
7   Tucker *et al.* (1986). (*Lactuca* xylem differentiation)
8   Brandes & Kende (1968), Grayburn *et al.* (1982). (Bud initiation)
9   Naf (1979). (Fern prothalli)
10 Ballade (1970), Bonnett & Torrey (1965), Wochok & Sussex (1976). (Root/shoot conversion)
11 Basile & Basile (1983). (Leaf induction in liverworts)
12 Battey & Lyndon (1988), Cutter (1954), Snow & Snow (1933), Steeves (1966). (Leaf determination)
13 Deschamp & Cooke (1984), Sachs (1969). (Leaf form)

CHAPTER ELEVEN

Competence and determination in flowering

11.1 THE TRANSITION TO FLOWERING[1]

When the higher plant changes from vegetative to reproductive growth, a new set of structures, the flower, is formed and represents the expression of genes not previously expressed. The formation of sepals, petals, stamens, and carpels occurs nowhere else in the plant and meiosis occurs only in the stamens and carpels. The **transition to flowering**, therefore, represents a change in determination of the shoot apical cells and their derivatives. In a vegetative plant already competent to flower (ripe-to-flower), only a change in determination would be required. A lack of competence to flower is most clearly seen in those plants that show a juvenile phase, i.e. a phase in which it is not possible to cause flowering directly (see Ch. 10, section 10.3.5.2). Remembering that competence may be defined as the ability to develop along a particular pathway when given the appropriate signal, plants that flower in response to **photoperiod**, in which a signal is received from the leaves, may be regarded as competent to flower. In such plants, the leaves are obviously competent to form the floral stimulus, but they may not be in some juvenile plants. Furthermore, it is often far from clear whether this signal from the leaves first has to make the apical meristem competent to flower or whether the meristem is already competent to flower but simply lacks the signal for determination. We can see that competence may therefore reside in the leaves or in the apex, or neither, or both (Fig. 11.1):

(1) The leaves are competent to provide a signal but the apex is not competent to react, i.e. juvenility resides in the apical meristem; this seems to be the case with meristems of young trees that remain juvenile even when grafted onto mature plants.

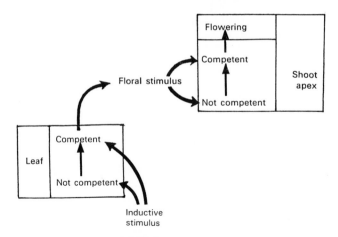

Figure 11.1 Possible changes in competence and determination required for flowering and the stages at which signals could act. The inductive stimulus may be required either to make the leaf competent to produce the floral stimulus or to switch on its production. At the shoot apex, the floral stimulus could be required either to make the apex competent to react to further stimulus or to initiate floral determination. In leaves and apex, the signal could have a dual role.

(2)   The leaves are not competent to provide a signal but the apex is competent to react; this is the case in *Bryophyllum*, in which the apex from a juvenile plant will form flowers if grafted to a mature plant.

(3)   The leaves are not competent to provide a signal, nor is the apex competent to react; this seems to be the case with plants requiring **vernalization** followed by a specific photoperiodic requirement. Vernalization (treatment at low temperatures of about 5°C for several weeks), which acts on the apex and not on the leaves, seems to change the competence of the apex to react to signals produced by the leaves in the appropriate photoperiod. Examples are the biennial henbane (*Hyoscyamus*), beetroot and other cold-requiring biennials, and winter cereals, which require vernalization followed by long days (LD) to flower.

(4)   Leaves and apex are both competent and the plants flower rapidly in response to an inductive signal.

Plants with a photoperiodic requirement to flower could fall into any one of these four categories. Plants that flower in response to a single photoperiod are likely to be those in which the apex is already competent to flower and requires a signal from the leaves for flower determination. *Pharbitis* is such a plant, but it does not respond to photoperiod until it has first been made competent to do so by a short exposure to light some time during the initial dark period, 10 min of red light being sufficient. Whether this makes the cotyledons competent to form the floral stimulus

233

Table 11.1 Effect of gibberellic acid (GA_3) on induction of flowering in *Bryo-phyllum*. *Bryophyllum* requires long days (LD) and short days (SD), in that order, to induce flowering. This implies that the LD make the leaves competent to react to the SD. GA_3 can substitute for LD in inducing competence of the leaves. (Zeevaart 1969)

	Flowering response	
Treatment	$-GA_3$	$+GA_3$
SD	−	+++
LD	−	−
LD → SD	+++	+++
LD → SD → LD	+	++
SD → LD	−	±

or makes the apex competent to react is not known. *Bryophyllum* is a plant having (when mature) a dual requirement (LD followed by SD) in which LD can be replaced by gibberellin (GA) (Table 11.1). This can be interpreted as showing that GA causes the leaves to become competent to respond when they are given short days (SD). Some plants with dual photoperiodic requirements and some of those that require repeated photoperiodic cycles in order to flower could be those in which competence needs to be acquired by the apex. For instance, in the biennial henbane, vernalization can be replaced by GA, which alters the competence of the plant to respond to subsequent LD. Plants of this sort seem to be those in which different environmental signals are required in order to make the apex competent to react and to induce the leaves to produce a floral stimulus.

Other plants may require a relatively prolonged environmental stimulus for flowering in order to first induce competence before determination can be induced. For instance, *Silene coeli-rosa* requires four or more LD in order to induce flowering. Events during the first three LD are essential for flowering, although these three LD by themselves are non-inductive. This could be because the competence of the apex to respond is changed during the first three LD and then substances already present or newly synthesized in the leaves are required to elicit determination of the apex to form the flower. The acquirement of competence must precede determination but if both can be responses to the same environmental signal then both would be part of the process of commitment of the apex to flowering.

The time of commitment of the apex to flower has been measured in some photoperiodically sensitive plants (*Sinapis*, *Lolium*, and *Xanthium*) by applying metabolic inhibitors to the apex at different times during and after the period of induction and finding out when it is they cease to be

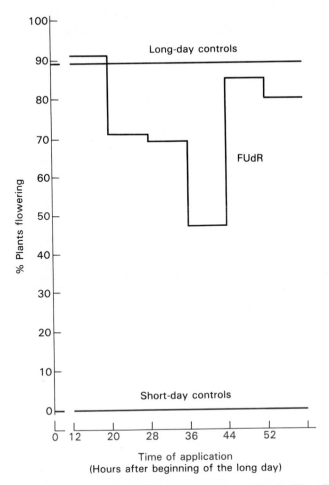

Figure 11.2   Determination for flowering in the shoot apex of *Sinapis*. When fluorodeoxy-uridine (FUdR, an inhibitor of DNA synthesis and growth) was applied to the apical bud, it inhibited flowering until 44 h after the beginning of induction, by which time the apex was presumably irreversibly determined for flowering. (Kinet *et al*. 1971)

effective in preventing flowering. In *Sinapis*, flowering can be prevented (and the plant will continue to grow vegetatively) up to 44 h from the beginning of the inductive long day (which lasts 0–18 h). After 44 h, flowering is inevitable. Commitment to flower, therefore, apparently occurs within a few hours, about 44 h from the beginning of induction (Fig. 11.2). In *Sinapis*, the leaves are, therefore, competent to respond to LD by producing a stimulus that reaches the apex about 13 h after the beginning of induction. The apparently competent apex then becomes irreversibly determined as floral over the next 31 h.

11.2   COMMITMENT TO FLOWER IN PLANTS WITH JUVENILE PHASES – TOBACCO AND BLACKCURRANT[2]

When the shoot apex of a vegetative tobacco plant that has already produced 20 or more leaves is cut off and rooted, this apical cutting will grow to form a plant of about normal size, producing many leaves before it flowers. The total number of leaves formed by the apical meristem right from its beginning until it flowers is almost doubled, but can be increased even more by repeating the excision and re-rooting the apical portion of the plant. In this way, flowering can be delayed until over 100 leaves have been produced instead of the usual 40 or so (Table 11.2). The plants can also be made to produce many more leaves than normal before flowering when root formation is induced well up the stem by surrounding that part of the stem with soil to induce roots there. The apex, therefore, only becomes competent to flower when it is sufficiently far from the roots. The plant behaves as though it were counting the number of leaves between the shoot apex and the roots. This implies that the roots in some way prevent determination of the apex for flowering.

When flowering does occur, only those axillary buds near the tip of the plant are able to form flowers and are, therefore, determined for flowering. This is tested for by cutting off the bud in question and rooting it so that it grows to form a new plant. The number of leaves formed by these rooted buds before they form flowers is compared with the number of leaves that they would have formed if left on the plant but made to grow out by decapitating the plant just above them. When excised and rooted, the fifth youngest axillary bud makes about the same number of leaves as a young plant, i.e. it goes back to the beginning, whereas the first and second youngest axillary buds make only as many leaves as they would have done had they remained on the plant (Fig. 11.3). This shows that the first and second buds are determined for flower formation, whereas the fifth bud is not. Determination for flowering, therefore, occurs at about the level of the third or fourth buds. The necessity for the apical meristem to be a minimum distance from the roots before determination for flowering can occur means that the roots are able to inhibit the processes of floral determination, probably by the production of some growth regulator.

One measure of competence to form flowers is the ability to form flower buds by pieces of tissue taken from various parts of the plant and cultured on an appropriate medium. These may be internode slices or epidermal **thin cell layer explants** (see Ch. 10, Fig. 10.5). Flowers are formed in culture only by tissues taken from plants already flowering. The ability of internodal tissues to form flowers in culture can extend further down the stem than the buds that are determined, suggesting that cells become competent for flowering several internodes before they

236

Table 11.2 Number of leaves produced by continually re-rooted buds of tobacco. The apical 5 cm or so of the plant was cut off and re-rooted each time the plant began to bolt. By preventing the shoot apex from becoming sufficiently far removed from the roots, this caused the production of many more leaves than normal and delayed determination for flowering. (After Dennin & McDaniel 1985)

Treatment	Number of times re-rooted	Total number of leaves produced
Controls (intact plants)	None	41
Apical 5 cm excised and re-rooted	3	97
	3	89
	5	93
	6	127

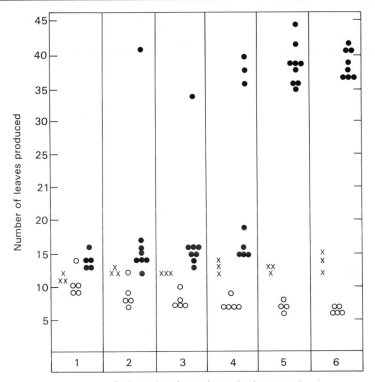

Figure 11.3 State of determination of buds on flowering tobacco plants. The youngest buds, when excised and rooted (●) and allowed to grow on, form only as many leaves before flowering as they would have done on the intact plant. The older buds form as many leaves as would have been formed by a plant grown from seed. The youngest buds are, therefore, determined for flowering but the older buds are not. (x) Number of leaves formed by the axillary buds when left on the plant. (o) Number of leaves present on the axillary buds at the time of rooting. (Dennin & McDaniel 1985)

become determined. However, this could be because the individual cells are competent, or that they become competent in culture in the absence of roots, or that the intact buds do not become competent until a specific stage in their development.

Experiments with blackcurrant (*Ribes*) have given similar results. Only shoots with >22 nodes, when floral induction began, gave 100% flowering. The initiation of **adventitious roots** half-way up 25-node plants (as in tobacco), before giving them six weeks of SD, prevented flowering in nine out of ten plants. Also, when large plants were decapitated, only those left with 27 or more nodes invariably flowered when induced by three weeks of SD. When shoots of mature plants were cut into sections and rooted, and then given six weeks of SD, they did not become induced to flower. However, if the pieces of stem were given SD before roots were formed and while rooting was suppressed, 60% did eventually flower. Again, the roots seemed to be preventing the floral determination of the shoot apex. Tests of the effects of applied plant growth regulators showed that gibberellic acid could prevent flowering totally in SD. Assays for gibberellin-like activity in plant extracts showed gibberellin to be present in roots and juvenile stems of the blackcurrant but not in the upper parts of adult stems. The conclusion is that, in *Ribes*, GAs from the roots appear to prevent floral determination in response to SD.

These experiments with tobacco and blackcurrant, and similar experiments with sunflower (*Helianthus*), are consistent with substances produced by the roots inhibiting the floral determination of the apical bud. It is not clear whether this is because they inhibit the acquirement of competence by the bud or whether they antagonize or neutralize the signals from the leaves that would cause the buds to become florally determined.

In *Pharbitis*, the axillary buds can react to the floral stimulus and form flowers only at a specific developmental stage when they have no more than two primordia (Fig. 11.4). Similar observations on *Scrophularia* explants showed that roots inhibited flower initiation and that the inhibitory effect of the roots could be replaced by kinetin, a cytokinin (Fig. 11.5). It is not entirely clear whether the kinetin is inhibiting the acquirement of competence of the buds or is preventing the action of a signal that would elicit determination in otherwise competent tissues. What such experiments do show, however, is that competence for floral determination may last only for a short time in some buds and may be a function of their stage of development. The new leaves formed on the buds themselves may inhibit floral determination, except when they are very small. Young leaves are potent inhibitors of flowering in the succulent, *Kleinia*. These several examples all show that the acquiring of competence by a meristem to form a flower may be inhibited by growth substances produced by the roots or the leaves.

238

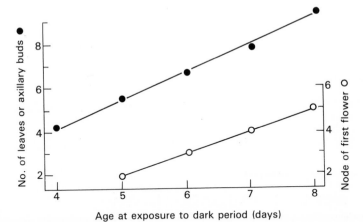

Figure 11.4 Competence of *Pharbitis* buds to form flowers. At whatever age plants were induced to flower by a 15 h dark period, the youngest axillary bud to form a flower was always the fourth below the apex. At this stage of its development, the bud had not yet formed its second primordium. Older buds with two or more primordia at the time of induction were already determined as vegetative. (King & Evans 1969)

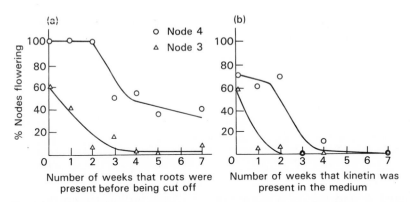

Figure 11.5 Effects of roots or cytokinin on the competence of *Scrophularia* explants to form flowers. Explants consisting of the stump of the stem with its attached roots and a cotyledonary bud present were cultured in agar. (a) The longer the roots were left on the explants before they were cut off, the greater was the inhibition of flowering. The lack of effect on the fourth node before Week 2 was because it was not initiated until then. (b) When explants without roots were given 10^{-6} M kinetin (a cytokinin), the longer the explants were supplied with cytokinin the greater was the inhibition of flowering. The cytokinin therefore mimicked the roots in inhibiting flowering. (Miginiac 1972)

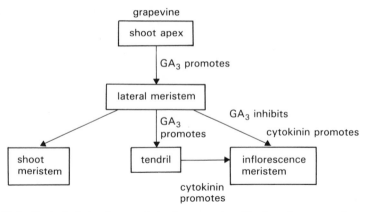

Figure 11.6 Control of morphogenesis in the grapevine. The lateral meristem has three potential fates. To bring about the formation of an inflorescence meristem, the gibberellic acid (GA_3) and cytokinin must be given in the correct sequence. (After Mullins 1980)

11.3 EFFECTS OF GROWTH SUBSTANCES ON DETERMINATION IN THE GRAPEVINE[3]

A good example of the ability of growth substances to control the determination of meristems is in the grapevine (*Vitis*) (Fig. 11.6). The shoot apex produces lateral meristems, which can develop into lateral shoot meristems, tendrils, or inflorescence meristems. Lateral meristems are formed only in the adult and not in the juvenile plant. In grape cultivation, the interest is in increasing or controlling the number of inflorescences and hence the number of bunches of grapes that are produced. Gibberellic acid (GA_3) applied to the shoot tips promotes the formation of lateral meristems. Continued application of GA_3 promotes their transformation into tendrils but if GA_3 is no longer applied then a lateral meristem will tend to develop into a vegetative shoot meristem. If, however, cytokinin is applied to a lateral meristem, this promotes its development into an inflorescence meristem. Thus, a lateral meristem is competent to develop in several ways. The formation of a lateral meristem and hence its competence to form further structures is promoted by GA_3. The determination of this structure as a tendril is promoted by GA_3 and its determination as an inflorescence meristem is promoted by cytokinin and inhibited by GA_3. If the order of application of GA_3 and cytokinin is reversed, then both are ineffective. Only the adult form of the grapevine is competent to form lateral meristems. By promoting their formation, GA_3 is, therefore, acting to reduce juvenility, the opposite of its action in ivy (see Ch. 10, section 10.3.5.2). A lateral meristem is competent to form an inflorescence but can be inhibited from doing so by GA_3 and promoted by cytokinin. Growth substances can clearly control changes in competence and determination for flowering in the grapevine. Whether they do so *in vivo* is not known but seems very probable.

11.4   DETERMINATION IN THE FLOWER[4]

When developing flowers of *Primula* at different stages during their development were bisected vertically, new primordia developed adjacent to the cut surface on the shoot apex. Normally, they would not have formed here because this was originally the centre of the apical dome. Only organ types that had not yet been formed at the time of the cut were formed on the cut meristem. When a cut was made at the sepal stage, for example, only petals, stamens, and carpels would form, never sepals. The conclusion was that the apex progressed through a series of physiological states (states of competence) that succeeded each other and allowed the apex to form only one kind of organ at any particular time. These experiments have been repeated with the same results for *Portulaca, Nicotiana*, and *Aquilegia*. They are consistent with the hypothesis that the flower meristem behaves as a relay system in which determination of one set of organs depends on the metabolic state of the meristem, which changes as successive sets of organs are formed. There is evidence that floral organs produce growth substances (Table 11.3) and so the substances produced by one set of organs could perhaps influence the determination of succeeding sets. The determination of the floral apical meristem is, like the determination of the leaf (see Ch. 10, section 10.8.3), progressive and sequential.

Growth substances can regulate the growth and size of various floral organs when applied to flowers *in situ* or excised and in culture. Gibberellins, and sometimes auxins, may promote flower development. Petals and stamens have been claimed to be sources of auxin, and stamens and ovaries may be sources of gibberellins. Growth substances may affect not only the growth of the floral organs but also their initiation and developmental pathways, i.e. their determination. The classic case is that of unisexual flowers in which normal stamens may be formed in otherwise female flowers under the influence of GA and femaleness may be promoted by auxins. The sex ratio in a number of species, e.g. spinach, can be controlled by the application of growth substances (Table 11.4). Flower anomalies, such as the formation of ovules on stamens, can sometimes be produced under the influence of growth substances. Aberrations of flower form can also sometimes be produced by photoperiod and so the effects of photoperiod may be mediated by plant growth regulators.

Table 11.3 Production of growth substances by floral organs and promotion of their growth. These results are merely a very broad summary. The production and effects of growth substances may change with the stage of development of the flower and its organs. Effects may also depend on the experimental conditions. Abscisic acid tends to delay development. Ethylene tends to promote femaleness, and GA (gibberellin) tends to promote maleness (see Table 11.4). (After Kinet *et al.* 1985)

Floral organ	Growth substances produced	Growth substances promoting growth
Sepals	GA	GA Auxin
Petals	GA	GA Auxin
Stamens	Auxin GA	GA Auxin
Ovary	Auxin GA	GA Auxin
Whole flower		Cytokinin

Table 11.4 Control of flower sexuality in spinach. The male and female flowers are borne on separate plants. Growth substances were applied via the roots when the plants had developed three leaves. GA_3, gibberellic acid; 6-BAP, 6-benzylaminopurine; IAA, indole acetic acid; ABA, abscisic acid. In spinach, gibberellin promotes maleness and auxin, cytokinin, and abscisic acid promote femaleness. (Chailakhyan & Khryanin 1980)

Treatment	Percentages of plants		
	Male	Female	Intersexes
untreated	48	52	–
GA_3 (gibberellin)	79	16	5
6-BAP (cytokinin)	11	87	2
IAA (auxin)	21	76	3
ABA (abscisin)	29	71	–

11.5 DETERMINATION IN THE FLOWER AS REVEALED BY REVERSION[5]

Flowering is normally the end of the line for a meristem. Even when the flowers are formed from lateral buds, the terminal meristem of the inflorescence eventually either differentiates into a flower or it senesces and dies before it gets to this point. In some plants, of which the pineapple is the best-known example, reversion of the meristem to vegetative growth occurs naturally, producing the characteristic tuft of leaves at the top of the inflorescence and fruit.

In plants with a terminal inflorescence, flowering normally continues until a terminal flower is formed. However, in some species, if the

Figure 11.7 Flower reversion in *Anagallis*. When plants are returned to short days after flower induction in long days, some flowers may show reversion to vegetative growth. When the flower meristem reverts to forming leaves at the stage of carpel formation, a leafy stem forms in the centre of the flower. (Brulfert 1965)

flowering plant is transferred to unfavourable conditions, eventually aberrant flowers are sometimes formed and by stages the apex reverts to producing leaves. This is floral reversion, seen most clearly in those few plants in which the flowers themselves revert to vegetative growth, with the result that a flower may grow on to form, at its centre, a stem bearing leaves, as in *Anagallis* (Fig. 11.7).

Impatiens is particularly interesting because only in this plant has it been possible to produce reversion consistently enough to allow experiments on reverting flowers to be carried out. Reversion of the terminal flower can be produced to order in *Impatiens* by transferring the plants from inductive SD to non-inductive LD. Leaf production can then be resumed at any time during flower initiation, according to the number of SD given. This is because the shoot apex does not become determined for flower formation. For the apical organs to be determined as floral parts rather than as leaves, it seems that a floral promoter must be produced continuously by the leaves in SD. Alternatively, a promoter of leaf determination might be produced in LD. Whichever is correct, the implication is that the apex itself does not become committed to the flowering state. The determined flowering state is not self-perpetuating in the *Impatiens* shoot apex as it is in plants in which induced leaves can be excised once flowering has begun, and yet flowering continues, e.g. as in *Sinapis* and *Lolium*.

Different reversion types can be produced by *Impatiens* according to the number of SD given. In one reversion type, organs intermediate between leaves and petals are formed, being leaf-like but having limited areas of petal pigment, the smallest of which are about 1–5% of the organ surface (Fig. 11.8). This indicates that the leaf surface consists of about 20–100 unit areas, which can each be determined independently. Since the area of an intermediate organ at initiation on the apical surface is equivalent to about 20–40 cells, determination as petal or leaf apparently occurs at, or very soon after, initiation and in each cell individually or in the clones derived from them. The earlier the SD are given, the greater is the proportion of each intermediate organ that develops a petal-like struc-ture. The probability of each cell or its derived clone forming a petal rather than a leaf structure, therefore, increases with the increasing exposure to SD while the cells are competent to respond. Because the tip of the organ becomes determined earlier than the base, the petaloid areas are most frequent at the base. This corresponds to the gradient of maturation found in the typical leaf (see Ch. 8, Fig. 8.6). Commitment becomes less and less reversible until eventually the whole primordium is determined when the primordia are about 750 μm long, about 11 days after initiation.

Competence to react to SD is increased by previous exposure to SD in *Impatiens*. Plants that have had five SD before a period of LD react quicker to a new dose of SD. The initial dose of SD acts only on new primordia as they are initiated, but the second dose of SD also affects primordia already present on the apex and diverts their development. These primordia are, therefore, already competent to react. The first prim-ordium recognized as a petal (>50% area pigmented) after continuous SD is initiated five days after the beginning of the SD treatment, whereas the comparable primordium after 5SD + 9LD + 5SD is initiated about two days *before* the beginning of the second period of five SD. This prim-ordium, therefore, responds to the second treatment of five SD about seven days sooner than it would have done to the first set of SD. Since it is already competent to respond, then seven days may be taken as the time normally required for the primordia to acquire competence.

11.6 DETERMINATION IN ABNORMAL FLOWER DEVELOPMENT[6]

Abnormal flowers turn up from time to time as monstrosities or as **homeotic mutants**. Abnormalities like double petals or the wrong number of floral parts are so common as hardly to qualify as abnormal. Less often, abnormal floral organs are formed, such as the intermediates

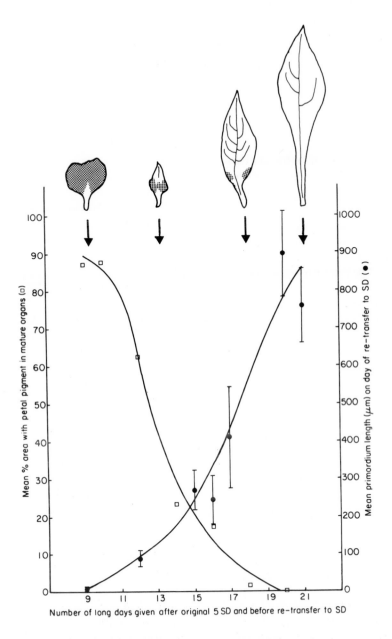

Figure 11.8 Distribution of petal pigment on intermediate organs in *Impatiens* and progressive determination of primordia. A primordium initiated when the plants are returned for re-induction in a second period in short days (SD) develops as a petal. The longer the period in long days (LD) before re-induction in SD, the more the primordium has become determined as a leaf. Pigmented areas are the shaded areas in the mature organs. (Battey & Lyndon 1988)

between leaves and petals in *Impatiens*, or petals bearing anthers, or ovaries bearing stamens or having flower buds within them. There is a tomato mutant in which the stamens are green, non-functional, and bear external ovules. The normal stamen phenotype can be restored by application of gibberellic acid (GA_3) to the bud at the time of stamen initiation. If the treatment with GA_3 is delayed, its effect is progressively less and less until it is without effect when the stamens are 0.8 mm long. These experiments show that:

(1) GA_3 can control the pathway of determination;
(2) determination as it progresses can stabilize at any point along the developmental pathway;
(3) the effect of the growth substance depends on the competence of the cells to respond, which is inversely related to the degree of determination achieved.

The many bizarre abnormalities that have been found show that there can be a complete gradation from leaves through all types of floral organs. In unisexual plants, one or more of the types of floral organs can be omitted altogether. Some spontaneously originating anomalous flowers of *Silene* were of particular interest because they showed, in a single flower head, the whole range of ovary structure from an open leaf with ovules on its margin to a normal closed ovary with free-central placentation at the centre of the 'flower'. This shows that there can, exceptionally, be a complete gradation in structure from leaf to ovary in the same plant and implies that, in this case, the processes of ovary determination could be halted at any point before determination had reached normal completion. This, in turn, implies that the genes for the ovary were expressed to a quantitatively greater extent as the meristem developed. Normal flower formation can be regarded as essentially a continual change in competence, determination, and gene expression at the shoot apex in which the different floral organs are singularities in this continuum.

11.7 FLOWERING AS CHANGES IN COMPETENCE AND DETERMINATION

Flowering depends on the leaves being competent to produce substances that promote flowering or to stop producing inhibitors, and being triggered, if necessary, to do so by environmental signals. Flowering also depends on the shoot apex being competent to flower, then receiving the appropriate stimulus or no longer receiving inhibitors from elsewhere in the plant, and on the determination of the floral organs in the appropriate sequence. This seems to depend on successive changes in competence of the apical meristem. Flowering is a process in which there are many changes in competence and in which the processes of determination at the apex change regularly so that a succession of different end states, the floral organs, is produced. Again, what we know of changes in competence and determination in the shoot apex is at least consistent with the involvement of plant growth substances as regulators. But, again, the problem is how the regulators themselves are regulated.

The floral stimulus produced by the leaves as a result of induction is usually thought of as a single substance which causes flowering when it reaches the shoot apex. Because the stimulus is graft-transmissible between different species (but only within the same family), the hypothesis of a floral stimulus common to related species has been extrapolated to that of a single substance that universally promotes flowering. Clearly this is not so. Gibberellins can act as floral promoters in some conifers and many long-day rosette plants. Auxins or ethylene trigger flowering in the Bromeliads (e.g. the pineapple). We have seen that in grapevines only a particular sequence of substances is effective. We need to consider flowering not as a single event requiring a single unique triggering signal but as the result of changes in the competence of the leaves to produce substances which, in turn, modify the competence and determination of the apex (Fig. 11.1). It would then be rational to regard the floral stimulus as the sequence of substances, or changes in the concentration of one or more substances, that cause changes in competence and determination at the shoot apex. In some callus, habituation (see Ch. 10, Box 10.1) for auxin can be brought about by changes in either cytokinin or auxin concentration. In *Convolvulus* leaf explants (see Ch. 10, Fig. 10.2), only the induction of determination seemed to require a specific substance, whereas acquirement of competence and the development that revealed determination had less specific requirements. Similarly, in flowering we might expect that in any group of closely related species there may perhaps be one step in the process that requires a specific stimulus but that other steps may be less specific, although a particular sequence of inductive changes may be necessary. The universal flowering hormone has never been found because it almost certainly does not exist and there seems to be no good reason why it should.

11.8 SUMMARY

(1) For flowering to occur, the leaves must be competent to produce any stimulus that is required by the apex and then must be able to produce it autonomously or be triggered to do so. The shoot apex must also be competent to respond and must then become progressively determined as the successive floral organs are formed. Juveni-

lity may depend either on the leaves not being competent to produce a floral stimulus or on the apex not being competent to respond. This can be tested by grafting apices from juvenile plants onto mature stocks. If the apex then flowers, juvenility would seem to be controlled by the leaves; if not, then the apex itself would appear to be juvenile and not competent to flower.

(2) In tobacco, the apex does not become determined for flowering until the plant has grown big enough for the apex to be sufficiently distant from the influence of the roots. Determination for flowering occurs in about the third bud below the apex. In blackcurrant, the apex is also prevented from becoming determined by substances from the roots, apparently gibberellins. In *Scrophularia*, cytokinins produced by the roots appear to be the inhibitors of apical determination. In the grapevine, competence and determination for lateral meristem formation and flowering in mature plants can be controlled by gibberellin and cytokinin. In order to produce inflorescence meristems, the growth substances must be applied in the correct sequence of GA then cytokinin.

(3) Surgical experiments on developing flowers have shown that the apex seems to pass through a succession of stages of competence as the successive types of floral organs are formed. Growth substances produced by the floral organs may be involved in regulating competence and determination within the flower. In a tomato mutant that forms aberrant stamens, gibberellin can restore the normal form of the stamen and acts to modify determination, its effect being progressively less as the stamens become more determined. In plants showing reversion of the flower to vegetative growth, determination of the floral organs seems to be under the control of substances produced by the leaves and is not under the autonomous control of the apex as it is in most plants. Reverting flowers can form organs intermediate between types, and organs that are partly one type and partly another. This shows that, in these cases, determination is probably at the level of the individual cells or their immediate descendants and not at the whole organ level. These and other anomalies suggest that there can be a continuous gradation of gene expression from vegetative to flowering, and even from one organ type to the next.

(4) When flowering is considered as a succession of stages of competence and determination, it seems plausible that these could be under the same sort of control by growth substances as other cases of determination may be. What seems to be common to different plants is not the substances that trigger competence and determination but the nature of these processes themselves, of which so little is known.

FURTHER READING

Battey, N. H. & R. F. Lyndon 1990. Reversion of flowering. *Botanical Review* **56**, (in press).

Bernier, G. 1988. The control of floral evocation and morphogenesis. *Annual Review of Plant Physiology* **39**, 175–219.

Bernier, G., J.-M. Kinet & R. M. Sachs 1981. *The Physiology of Flowering*, Vol. II. Boca Raton: CRC Press.

Francis, D. 1987. Effects of light on cell division in the shoot meristem during floral evocation. In *Manipulation of Flowering*, J. G. Atherton (ed.), 289–300. London: Butterworth. (*Silene*)

Kinet, J.-M., R. M. Sachs & G. Bernier 1985. *The Physiology of Flowering*, Vol. III. Boca Raton: CRC Press. (Growth substances and correlative growth in flowers)

McDaniel, C. N., S. R. Singer, J. S. Gebhardt & K. A. Dennin 1987. Floral determination: a critical process in meristem ontogeny. In *Manipulation of Flowering*, J. G. Atherton (ed.), 109–120. London: Butterworth.

NOTES

1 Kinet *et al.* (1971), Zeevaart (1969). (Commitment and competence of shoot apex)
2 Dennin & McDaniel (1985), King & Evans (1969), Kulkarni & Schwabe (1984), Miginiac (1972), Schwabe & Al-Doori (1973). (Juvenility and determination for flowering)
3 Mullins (1980). (Control of flowering in grapevine)
4 Cusick (1956), Jensen (1971). (Surgical experiments on developing flowers)
5 Battey & Lyndon (1988), Brulfert (1965). (Flower reversion)
6 Bowman *et al.* (1989), Komaki *et al.* (1988) (Homeotic mutants); Sawhney & Greyson (1979) (Tomato mutant).

PART VI

Coordination of development

CHAPTER TWELVE

Pattern formation, positional information, and integration of growth

12.1 SPACING PATTERNS OF ORGANS: PHYLLOTAXIS, LATERAL ROOTS, BRANCHING

The existence of patterns of cells, tissues, and organs implies coordination of the development of the units making up the patterns. There are basically two types of pattern-forming process. The order may be the result of an assemblage of similar but independent units like tiles, each with its own pattern, pushed together on a board. Alternatively, the pattern may be because of interaction between developing units, which promote or inhibit each other's development and so modify the pattern as it forms (differentiation-dependent pattern formation). Are all patterns in plants formed interactively and, if so, what is the nature of the interactions between cells, tissues, and organs, and how do they control plant form?

12.1.1 Leaves and phyllotaxis

When the positioning of new leaf primordia was examined (see Ch. 3), only the example of spiral phyllotaxis was considered but not the question of why the phyllotaxis was spiral rather than whorled, opposite, or distichous. F. J. Richards showed how spiral phyllotaxis could originate in the dicotyledonous seedling with two opposite cotyledons if a new primordium originated as far as possible from the previous primordia, and that these inhibited the initiation of new primordia close to them (see Ch. 3, section 3.2.2). If primordia are initiated one at a time, then the positioning of successive primordia would tend to an arrangement with a divergence angle approaching the Fibonacci angle of 137.5° (Fig. 12.1).

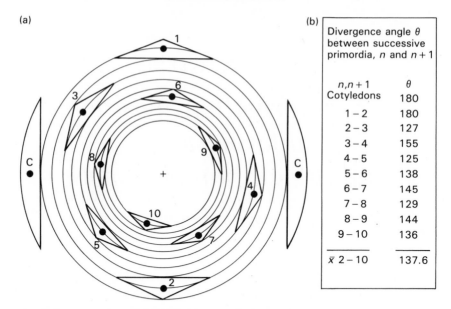

(a)

(b)

Divergence angle θ between successive primordia, n and $n+1$	
$n, n+1$	θ
Cotyledons	180
1 – 2	180
2 – 3	127
3 – 4	155
4 – 5	125
5 – 6	138
6 – 7	145
7 – 8	129
8 – 9	144
9 – 10	136
$\bar{x}$ 2 – 10	137.6

Figure 12.1 The establishment of spiral phyllotaxis in a dicotyledonous seedling. (a) The two cotyledons (C) are opposite each other (at 180°). Each new leaf primordium is positioned: (i) to be as far away as possible from the previous primordium; (ii) with respect to the then adjacent youngest pair of primordia, to be further away from the younger than the older in the inverse ratio of their age in plastochrons since their initiation. These 'rules' would be consistent with each new primordium producing an inhibitor of primordial initiation, which decreases in effectiveness as the primordium ages. This model results in the third primordium being the first to be not orthogonal to the cotyledons. A left-handed spiral is initiated. Equally probable would be a right-handed spiral. (b) The mean divergence angle between successive primordia (2–10) is close to the Fibonacci angle (137.5°). The beginnings of the spiral parastichies can be discerned: three in one direction (primordia 8,5,2; 9,6,3; 10,7,4) and five in the other (primordia 6,1; 9,4; 7,2; 10,5; 8,3).

The problem is, therefore, not so much how a spiral leaf arrangement might arise but how the very precise geometrical arrangements are arrived at, where the leaves are distichous (180° to each other) or opposite and decussate (successive leaf pairs being at 90° from each other and the leaves of each pair being exactly opposed at 180°). A morphogen model, which depends only on the production of an inhibitor of primordium initiation by the existing primordia (Fig. 12.1), does not seem entirely adequate to account for these precise arrangements because a small difference between the members of a pair in their inhibitory properties should tend to lead eventually to some spiral arrangement. However, the diffusion-reaction models of Meinhardt (see Fig. 12.15), in which an activator and an inhibitor are equally involved, can simulate opposite and distichous phyllotaxis very precisely. Also, the biophysical constraints imposed by the structure of the apical surface, as shown by Green (see Ch. 4, section 4.4), would seem to be able to provide the necessary precise framework for the maintenance (though not perhaps the origin) of a geometrical pattern, since this is prescribed by the existing pattern. The positioning of leaves may well be an example of patterning by two systems working in conjunction with each other. The morphogen theories tend to emphasize the mutually inhibitory nature of new primordia, whereas the biophysical theory emphasizes the dependence of new primordia on the existing structure.

The main example of change from spiral to geometric primordium arrangement is on formation of the flower. Does this imply a reduced role for the morphogen component and an increased importance of the biophysical component? Certainly, the positioning of primordia in the flower is not easily interpretable in terms of inhibitors produced by the primordia. It is much more understandable as the promotion of primordial formation by existing primordia, so that the pattern established by the sepals is maintained. If auxins are indeed involved as morphogens in primordial positioning in the vegetative plant (see Ch. 3, section 3.2.2), the general view that amounts of auxin decrease in the plant on flowering could be consistent with a lessening of the role of the morphogen component and an increasing role for the biophysical component.

According to the inhibitory field theory, the pattern would depend on the characteristics of the diffusible morphogens. Genes altering the physical characteristics of the morphogen molecules – promoters or inhibitors – would alter the phyllotactic pattern. Genes altering the rate of production or consumption of the morphogens would also alter the phyllotactic pattern. Perhaps the constant features of phyllotaxis – the leaf arrangement and the flower part arrangement, which are essentially invariable – could be programmed by the physical characteristics of the morphogens, i.e. the specific morphogens could be genetically defined. However, the changes in arrangement as the apex enlarges or when the plant flowers could be the result of changes in the metabolism of constant morphogens.

The biophysical and morphogen concepts both depend on there being a mechanism that allows or causes the apical surface to become deformed locally so that it can bulge out. If a morphogen produced at the site of primordium initiation is a substance that causes increased plasticity of the apical surface, it could interact with the surface structure, which may also be disposed in such a way as to allow or even to cause local deformation of the surface. Both systems could be reinforcing each other to cause primordium formation. If primordial positioning is a process governed by two cooperating processes, then if one is severely deranged perhaps the other can cope to allow normal or nearly normal development to continue. This could be an example of a built-in fail-safe system in the plant's development.

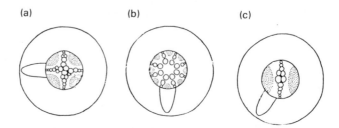

Figure 12.2 Radial position of lateral roots. Transverse sections show that lateral roots form: (a) usually opposite the xylem poles, (b) in some species opposite the phloem (shaded), and (c) occasionally opposite the junction of xylem and phloem. (Esau 1965)

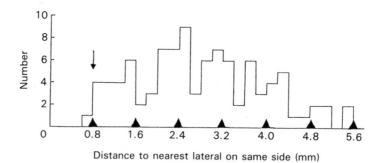

Distance to nearest lateral on same side (mm)

Figure 12.3 Longitudinal spacing of lateral roots in tomato. The frequency distribution of interlateral distances shows spacings at multiples of approximately 0.8 mm, which is the minimal interlateral distance (↓). (Barlow & Adam 1988)

12.1.2 Lateral roots[1]

Lateral roots are positioned radially in relation to the vascular structure (Fig. 12.2) and so form longitudinal rows along the root. Their positioning relative to the vascular tissues suggests that something diffuses out of the vascular tissues either to promote lateral initiation directly, or indirectly by delaying maturation of some of the pericyclic cells so that they remain capable of initiating lateral roots. The spacing within each row of lateral roots is usually fairly regular, as in the tomato (Fig. 12.3). The different rows in any one root usually do not seem to interact with each other. This would be consistent with the factors determining longitudinal spacing being transmitted down the vascular system.

In tomato, there are only two longitudinal rows of lateral roots, corresponding to the two opposite poles of the diarch xylem. Lateral root primordia therefore form, as seen in transverse section, at the 12 o'clock and 6 o'clock positions. But this accounts for only one-third of the roots. Another one-third form at 11 o'clock or 5 o'clock and another one-third at 1 o'clock or 7 o'clock. To account for this pattern, Barlow & Adam

256

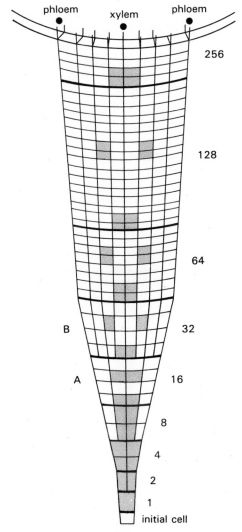

Figure 12.4 A possible cell lineage mechanism to account for spacing of lateral roots in tomato. At each round of cell division (e.g. A → B represents one round of cell division from 16 to 32 cells), the potential to form laterals (shaded cells) is partitioned to only some of the daughter cells. Alternative polarities in different ranks of cells at the 32-cell stage could account for the observed pattern. (Barlow & Adam 1988)

proposed a cell lineage mechanism for root initiation (Fig. 12.4). If cell divisions are regular and lateral-root mother cells are formed originally from alternate cells in a file (A in Fig. 12.4), comparable to unequal divisions, and can be either adjacent or separated derivatives of a subsequent division (B in Fig. 12.4), then the observed spacing could be accounted for.

When the longitudinal spacing can be decreased, e.g. by inducing more roots with auxin, as in *Raphanus* (see Ch. 3, Fig. 3.1), it is not clear whether the auxin is in some way increasing the nutrient supply to overcome a local inhibitory effect produced by existing roots using up the nutrients in their immediate vicinity, or whether the auxin is stimulating more pericyclic cells to divide to influence a cell-lineage patterning mechanism. Pattern derived from cell lineage may, therefore, be able to account for the detailed positioning of roots in relation to the vascular tissues and may also be able to account for the basic longitudinal spacing patterns. So far, there is no direct information on cell lineages during lateral root initiation.

12.1.3   Branching

12.1.3.1   Branching in *Griffithsia*[2]

The small red seaweed *Griffithsia* consists of a branched filament of shoot cells, each multinucleate and up to 1 mm long and, at the base, narrower, longer rhizoidal cells (Fig. 12.5). When a shoot cell is isolated from the rest of the plant by cutting the cells adjacent to it, it regenerates a plant in a predictable and regular fashion. A new shoot cell is cut off by an unequal cell division at the apex and a rhizoidal cell by a cell division at the base. The polarity of the isolated cell has, therefore, been maintained. New shoot cells are added (as in the intact plant) each by an unequal division of the apical cell. When the filament is 4–5 cells long, branch formation may begin by budding from the top of the second or third cell from the apex, the bud then becoming separated off by a cell division. The branch cell then becomes the apical cell of a new branch filament. New branches form always acropetal to existing branches and usually 2–3 cells below the apical cell. Not every shoot cell branches so that successive branches are separated by non-nodal cells. Rhizoid formation at the base is similar but branching is rarer. Subapical cells, except when budding to form a branch, do not divide except when they are induced to renew division by being excised and isolated. This implies that the apical cell(s) may inhibit division in the rest of the filament.

Light is necessary since cell division stops in continued darkness. Increasing the irradiance above a low level does not affect the rate of division but it does increase branching and so increases the total cell number in the plant. Branching and the number of non-nodal cells may, therefore, depend on the amount of photosynthate available over and above that required for the growth and division of the apical cell, but what the mechanism is by which the apical cell exerts its dominance is not known. The fact that an isolated cell can produce the whole plant shows that pattern can be decided locally. In this, *Griffithsia* resembles trees, in which the branching pattern also seems to depend only on local

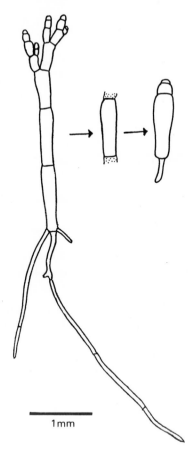

Figure 12.5 Branching and regeneration in *Griffithsia*. In this small red alga (with long basal rhizoids), the shoot branches by budding of a subapical cell. Non-nodal cells will begin asymmetric cell divisions at each end if the cell is isolated, and a new plant regenerates. (After Waaland 1984)

information. Again, the localized, modular construction of the plant and the importance of polarity is apparent, as is also the influence of the environment on plant form.

12.1.3.2 Branching in shoots and trees[3]
Branching can be regarded as a local program by which an axis divides. It can easily be shown to be local by severing a branch tip and growing it on – the same branching pattern will be maintained. Computer models of branching patterns can be produced by simple rules that essentially constitute a lineage mechanism if each local unit grows without reference to other units (Fig. 12.6a). In this particular model, successive branches are on opposite sides of the axis, which implies that the position of a

259

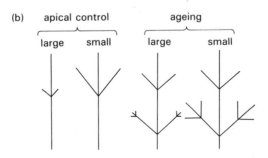

Figure 12.6 Models of branching. (a) A branching system generated by simple rules: A → I(B)C, B → A, C → I[D]A, D → C, I → I. Each branch generates branches at the same rate and there is a delay of one time unit before an apex forms a branch (i.e. B → A and D → C). (b) Branch elongation relative to the main stem can be influenced by the main stem (apical control), or by a requirement for ageing before growth. (Lindenmayer 1982 [a]; Cannell 1974 [b])

branch is influenced by the position of the previous one. Identifiable factors affecting branching pattern in the living plant can also include apical control and ageing (Fig. 12.6b). In the model in Fig. 12.6a, branching is assumed to occur at regular intervals. In the plant, this could depend on:

(1) the competence of the axillary buds to grow out as a function of their age;
(2) the nutritional status of the bud;
(3) the distance from the terminal or other buds, which would imply some sort of apical dominance, presumably mediated by inhibitors produced by these other buds.

Modelling can be useful in establishing whether a lineage mechanism could account for the observed branching or whether an interactive mechanism is more probable. A. Lindenmayer showed that to simulate the branching of the inflorescence of *Mycelis* (a Composite) with minimal assumptions, it was necessary to assume an apical dominance effect. This tallies with experimental observations on related species that cutting off the top of the inflorescence stimulates growth below.

The angle that a branch makes with the stem can be assigned arbitrarily in programs but its biological basis is not understood. It could depend on the structure of the leaf axil, the branch shoot simply growing out at an angle predetermined by the position and shape of the axillary bud. Branch angle may also be determined by the response to gravity and light, and the times during branch and internode development that the growth of the branch is affected.

The frequency of branching, and branch arrangement, will be related to the leaf arrangement. Where leaves are produced singly, branches will also usually be produced singly. Where leaves are in pairs or whorls, there is the potential for more than one branch at a node. In conifers, very few axillary buds form compared to the number of leaves. Those that do grow out into branches seem to be evenly (but not randomly) dispersed over the apical surface. Their distribution would be consistent with each inhibiting the outgrowth of adjacent branches within a definable radius. Whether this is by a nutritional effect or by the production of an inhibitor is not known. This spacing of the branches is analogous to the spacing of new leaf primordia in phyllotaxis (see Ch. 3, section 3.2.2).

12.2   PATTERNS OF TISSUES – BLOCKING OUT OF TISSUE PATTERNS IN MERISTEMS: WHY DO SOME CELLS DIFFERENTIATE AND NOT THEIR NEIGHBOURS?[4]

The primary tissues in plants characteristically differentiate not as single blocks or regions, where all cells of a particular tissue are together, but as several strands or nodules or at several loci separated by cells that have not become obviously differentiated or have differentiated into unspecialized parenchyma. The discontinuous nature of plant tissue differentiation is most apparent in cross sections of young stems with discrete vascular bundles. The formation of strands (see Ch. 7, section 7.6) has been explained as being the result of the canalization of the inducing stimulus, such as auxin, because of its accelerated flux in the differentiating cells causing stimulus to flow into the induced strand from the surrounding cells. They therefore become deprived of stimulus and so do not differentiate. The lateral distance between several strands induced by the shoot apex or some other single source of inducer would then depend on the kinetics of the flux of stimulus in relation to the time taken for cell determination. Which particular cells originate the strands could be random, with their spacing being determined by the characteristics of the system.

A similar problem exists in the induction of nodules of vascular tissue in callus, but here there is no flux of inducer through the cells (see Ch. 7, section 7.3). Why do the cells differentiate as vascular nodules with

261

undifferentiated cells left between them? Bünning suggested that the cells competed for nutrients. If a cell or group of cells beginning to differentiate tended to use up the local supply of nutrients and also began to deplete the neighbouring cells, then these neighbouring cells would not be able to differentiate. As with the supposed canalization of inducer into strands, the utilization of inducer at particular loci would tend to prevent differentiation at neighbouring loci. For this mechanism to work would require that all cells were potentially inducible, that some differentiated faster than others, and that the determination process was reversible so that cells about to differentiate, but then deprived of inducer, would revert to being uninduced and so would not then differentiate. All of these conditions are consistent with what we know of the processes of determination (see Ch. 10).

The same hypothesis seems equally plausible for differentiation in the root. In the root, the vascular tissue tends to form as a single central mass rather than as discrete strands. However, within the xylem of the maize root, the metaxylem vessels differentiate only at certain radial locations (Fig. 12.7). The number of metaxylem strands originating at the maize root tip is correlated with the root diameter (Fig. 12.8). New metaxylem files are initiated where the root widens behind the tip. A new file is initiated and inserted only when the original files become more than 50 μm apart because of the lateral and radial growth of the root tip (Table 12.1).

In excised and cultured pea root tips, the number of xylem poles is also a function of the root tip diameter. Roots cultured in auxin solutions were wider and had more xylem poles than those in the absence of auxin. Removal of auxin caused a decrease in the diameter of the root and of the number of xylem poles. The effect of the auxin in inducing new xylem poles is probably indirect, through its effect on root diameter. A similar phenomenon occurs naturally. As roots age, they may become narrower and more slender and this may be accompanied by a reduction in the number of xylem poles. In stems, too, the number of vascular bundles is often a function of the diameter of the stem, although this may be because wider stems are from larger plants that bear more leaves above the point of sectioning, the vascular bundles representing the leaf traces. These patterns seem consistent with the hypothesis that the formation of new loci of differentiation depends on the cells being sufficiently removed from existing loci that are presumably exhausting some critical nutrients in their immediate area.

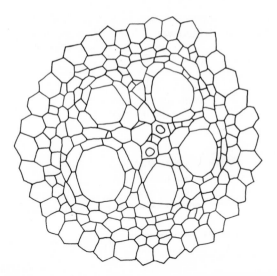

Figure 12.7 Pattern of metaxylem vessels in a maize root. In a transverse section of the stele, the metaxylem vessels are seen as the large cells more or less equidistantly spaced. (After Feldman 1977)

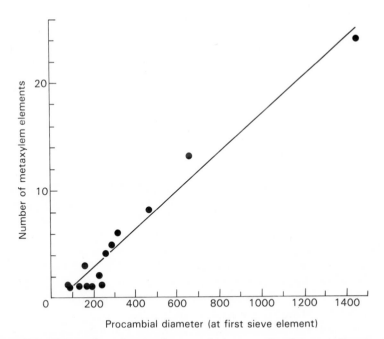

Figure 12.8 The number of metaxylem vessels is proportional to root diameter. The number of metaxylem vessels present in a maize root is linearly related to the diameter of the procambium, and this in turn is correlated with the diameter of the root. (Data of Barlow & Rathfelder 1984)

Table 12.1 Origin of metaxylem cell files in *Zea* roots. New files form in the region where the tip of the root widens, when the existing metaxylem files become more than about 50 μm apart. Numbers in parentheses show number of roots examined. (After Feldman 1977)

Number of metaxylem files	Distance (μm) behind root cap junction at which file originates (± SE)	Mean distance (μm) between existing files between which the new file originates
1–4	63 ± 2.5 (15)	42.5[a] (2)
5	78 ± 2.9 (15)	52.5 (2)
6	84 ± 4.9 (9)	59.5 (2)
7	93 ± 5.9 (7)	70.0 (1)
8	110 ± 11.1 (6)	–

[a] The mean distance between the first four files where they originate.

12.3 PATTERNS OF CELLS

12.3.1 *The spacing of stomata in the epidermis*[5]

The spacing of stomata in the epidermis of many dicotyledonous leaves is, at first sight, apparently random (Fig. 12.9a). The presence of younger stomata, at an early stage of development, in the spaces between more developed stomata led Bünning to propose that a stoma tended to inhibit the development of other stomata in its vicinity and this would lead to the stomata, as they developed and in the mature leaf, being more or less equally spaced over the epidermis. This hypothesis is a field theory, like that proposed for the initiation of leaf primordia (see Ch. 3, section 3.2.2), and implies that new centres of differentiation arise at minima that form in an inhibitory field produced by existing centres as they grow apart.

Stomatal and epidermal development has been examined by T. Sachs to see whether this theory is sustainable. When the spacing pattern is analysed by plotting the frequency of stomata as a function of distance from a given stoma, and many such measurements are superimposed, then it becomes apparent that there is a stoma-free area around each stoma corresponding to one or a few epidermal cells (Fig. 12.10). Beyond this region, the frequency of stomata seems to be a matter of chance. Is this stoma-free area the result of the existing stoma producing a short-range inhibitor of stomatal initiation? When the development of stomata was examined, the same result was found in a number of species. The pattern arose because of the polarized division of a cell to give a stoma mother cell, which then divided again, sometimes several times, to form the stomatal apparatus. The stoma-free area around the stoma was a

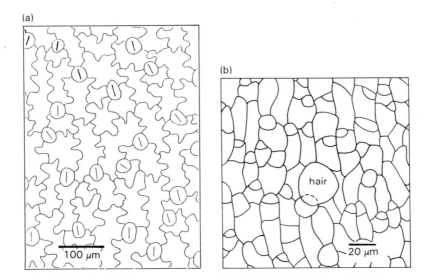

Figure 12.9 Stomatal spacing in *Anagallis* leaves. (a) Stomata are apparently randomly spaced. (b) This spacing arises because in early development many of the epidermal cells, which tend to be in a staggered arrangement, divide unequally but all with the same polarity so that the potential stomatal mother cells are separated from each other by non-stomatal cells. (Marx & Sachs 1977)

consequence of the original cells all being polarized along the same vector and each cell forming a stoma but only from the cell divisions at one end of the original polarized cell (Fig. 12.9b). Even when the cells are not polarized along a vector, divisions can result in the stoma itself being in the centre of a cell complex that necessarily separates it from neighbouring stomata (see Ch. 5, Fig. 5.7). The pattern, therefore, depends on the cell lineage and there is no need to postulate the production of inhibitors by the stomata. If the development of the stomatal complexes is not synchronous, then there can be a mixture of mature and developing stomata.

The dependence of stomatal spacing on cell lineage is often seen more easily in monocotyledonous leaves, as in *Crinum* (Fig. 12.11), where stomata are not contiguous because they develop from staggered, although contiguous, cells. Other factors apart from cell lineage may, however, also be involved in stomatal spacing. In those grasses that are C4 plants and show a Kranz-type structure, such as maize, the stomata are found only over the mesophyll and not over the regions of the vascular bundles. In many plants, the frequency of stomata is different over vascular and interveinal tissues. This seems to imply an influence of the underlying cells on the development of stomata, but the extent and nature of this influence has not been investigated. There are also many

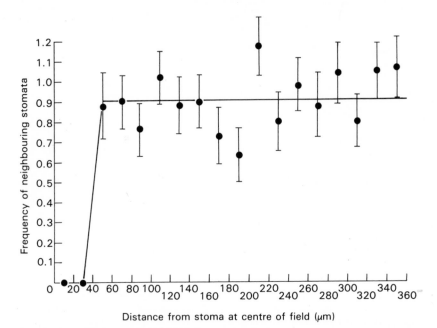

Figure 12.10   Analysis of stomatal distribution in the lower epidermis of *Anagallis* leaves. The frequency of occurrence of neighbouring stomata in a unit area (1600 μm^2) at various distances (normal to the stomatal long axis) from given randomly selected stomata. There is a stoma-free area around each stoma, which corresponds to the width of one epidermal cell. Similar graphs were obtained for measurements parallel to the stomatal axes, and for the upper epidermis. The horizontal line represents the expected stomatal frequency if this is independent of the distance to neighbouring stomata. (Marx & Sachs 1977)

instances of stomatal frequency being quite different on different surfaces of the same leaf, and this implies that there are controls on stomatal frequency other than cell lineage alone. In pea leaves, abortion of stomata developing close to other stomata implies interactions, the nature of which is unknown.

12.3.2   The spacing of hairs in the epidermis[6]

The spacing of hairs has many of the characteristics of the spacing of stomata, which is hardly surprising since both are modifications of epidermal cells. There seem to be basically two types of spacing of hairs: that due to cell lineage and that due to the positions of the underlying cells.

In grasses and some other plants, a potential root hair cell, a trichoblast, is formed as the result of an unequal division of an epidermal cell. According to the species, the small, densely cytoplasmic trichoblast cell may be either the proximal or distal derivative of the unequal division. In

(a)

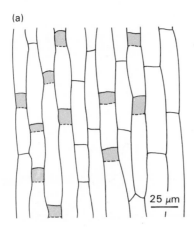

25 μm

(b)

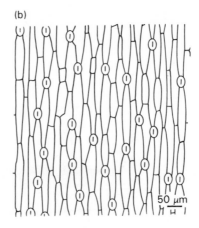

50 μm

Figure 12.11 Stomatal distribution in leaves of the monocotyledon *Crinum*. (a) In early development, epidermal cells divide unequally and all with the same polarity. The small cells (shaded) divide to form the stomata. (b) The cell lineage mechanism generates a regular spacing of mature stomata. (Sachs 1974)

the water plant *Hydrocharis*, the large distal epidermal cell divides several more times so that each trichoblast is separated from the next one on the cell file by several non-hair cells. The pattern, therefore, seems to depend on cell lineage, although the details of the pattern and whether it is modified by other factors have not been examined. Trichoblasts may be separated from those in adjacent files or may be contiguous, depending on the cell pattern at the time of the unequal division. Contiguous trichoblasts are consistent with a cell lineage spacing mechanism, but not mutual inhibition.

In *Sinapis* roots, hairs form only from those epidermal cells that lie over an intercellular space in the underlying cortical tissues. When the epidermis is peeled off, a new epidermis regenerates and hairs also regenerate, but again only over intercellular spaces in the underlying tissue. The pattern of hairs in the epidermis is, therefore, dictated by the structure of the internal tissue in which the cells are larger than the epidermal cells, so that only some of the epidermal cells are over spaces and so develop a hair. A possible mechanism is that underlying cells produce an inhibitor of hair formation and only those epidermal cells that are over spaces are sufficiently freed from the effects of the inhibitor for hairs to develop. This would imply that in isolated epidermis all cells should form hairs. This, in fact, is what happened when the epidermis and outer tissues of a *Sinapis* root were separated from the inner tissues by an oblique cut (Fig. 12.12). This seems to imply that the inner cells produce some inhibitory substance that normally passes to the sub-epidermal cells and prevents hair formation in the epidermal cells with

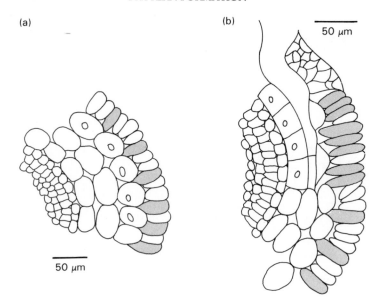

(a)

(b)

50 μm

50 μm

Figure 12.12 Hair formation in *Sinapis* roots. (a) In transverse section, hair cells (shaded) are seen only over the radial walls of the underlying cortical cells. (b) When the epidermis is separated from the inner tissues by an oblique cut, many more epidermal cells form hairs. (After Barlow 1984)

which they are in direct contact. As with stomata, spacing patterns can, therefore, depend on either cell lineages or on the position (and the influence) of neighbouring or underlying cells. The physiological basis of these interactions awaits investigation.

12.4   REGENERATION OF PATTERN – POSITIONAL INFOR-MATION

Experiments on bristle patterns and limb regeneration in animals have led to theories about positional information and its interpretation by cells in development. There are two principal theories, which may be called the gradient model and the polar coordinate model. Wolpert's gradient model assumes that cells recognize their position on a gradient of morphogen and interpret the concentration of morphogen by developing in a specific way. When tissues are removed, a new gradient is set up, the cells at the boundary becoming specified as one end of the new gradient. Cells along the new, shortened gradient then grow and develop to restore the original structure. In the polar coordinate model of French, Bryant & Bryant, the position of a cell in an organ corresponds to one of a continuous set of polar and radial coordinates, which specify its radial

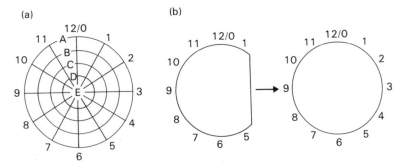

Figure 12.13 Polar coordinate model of positional information. (a) Each cell in an organ is assumed to have information with respect to its position on a radius (A – E) and on a circumference (0 – 12/0). (b) When cells are removed, wound healing results in adjacent cells having non-adjacent positional values. Cell division and growth then continues until the original coordinates are restored. The radial coordinates can be drawn upwards as they would be on a shoot apex, in which E would represent the apical summit and A the flanks. (After French *et al.* 1976)

and lineal position, respectively (Fig. 12.13). When tissues are removed, this necessarily means that the exposed tissues no longer have tissues with adjoining coordinates adjacent to them. The theory is that the cells then grow and develop until the original continuous coordinate system is restored. This model accounts more readily for regeneration of insect organs and tissues than the gradient model. Are these models applicable to plants? Or is Sach's model of differentiation-dependent development more appropriate? Are these necessarily mutually exclusive or could they coexist? Cell lineage-dependent pattern formation may be more characteristic of plants because their cells cannot move. Although unequal divisions are important in early animal development, they may be even more important in plants where cell polarity may also be more important because the cells do not move with respect to each other as they do in animals, where cell surface recognition systems may be more important.

12.4.1   *Severed and split apical meristems*[7]

Apical meristems that are cut off and cultured can produce the whole organ in culture. Indeed, isolated quiescent centres from roots can regenerate whole roots, maintaining their polarity as they do so, forming the root cap on the originally apical face and the body of the root from the basal face. This shows that the root initial cells are autonomous and do not depend on the presence of the other root cells for their continued functioning. Similarly, isolated shoot apical domes will grow in culture and form a whole shoot. Both root and shoot initial regions in culture require the addition of growth substances, such as auxins and cytokinins, to maintain their growth; in the intact plant, these growth substances are presumably supplied from other parts of the plant.

When parts of apical meristems are cut away, the missing parts can be regenerated by the cells that are left. The removal of half of the root cap causes the stimulation of cell division in the remaining parts of the root cap and in the quiescent centre. When the cap has been restored to its normal size, the rate of cell growth and division falls to its normal values and the normal growth of the root is resumed. How does the root sense that part is missing and that the regeneration process is complete so that regeneration growth then stops?

The same problems are raised by experiments with *Lupinus* in which the shoot apex was split vertically into six pieces, each of which could then regenerate a complete new apex after growth and cell division. Vascular tissues differentiated in the new apices, even those formed from tissues in which there was none originally. Some of the narrower apices formed a rod of procambium, but since the first leaf was formed to one side of the new apex the symmetrical procambium could not have been the determinant of the leaf position. Procambium also formed in the new apices above the level of the leaf. This did not happen in intact apices. In the larger regenerating apices, the procambium diameter was greater and could form a ring. This is comparable to the roots in which xylem diameter was a function of root diameter (Fig. 12.8) and suggests that procambium may form at some specific distance from the surface, as experiments with regenerating cambium imply (Fig. 12.14).

When a lupin shoot apex was isolated on a plug of pith tissue by four vertical cuts, a new apex was reconstituted. New epidermis and cortex were regenerated on the cut surface of the plug. A ring of cambium differentiated in what had been pith and the formation of leaf primordia was begun by the new apex. The leaves were formed in the normal spiral arrangement but were usually displaced from what had been the previous arrangement on the original, intact apex, so that a new phyllotactic system was established. Presumably, a new surface microstructure of the sort shown by Green (see Ch. 4, section 4.4.1) is established simultaneously with the new phyllotactic arrangement.

Other experiments in which shoot apices have been isolated have been done with the sunflower (*Helianthus annus*). Just as florets were about to be formed, the centre of the broad floral apex was isolated by a circular cut so that it formed a plug on the lower pith tissue. This plug began to form florets but not in the normal arrangement. It initiated its own, new phyllotactic arrangement. Florets first formed at one side of the plug and floret initiation then spread to the rest of the plug. Isolation therefore altered the arrangement and spatial sequence of the initiating florets, which was, therefore, not predetermined by the surface structure nor did it seem to depend on a mutual inhibition mechanism (see Ch. 3, section

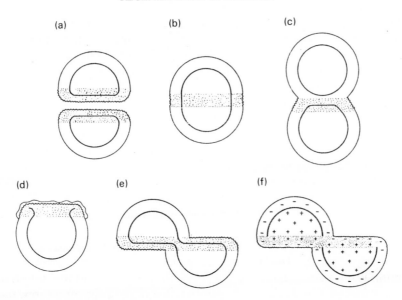

Figure 12.14 Regeneration of cambium. (a) In a split stem, cambium forms in the shaded area to link up the original (severed) cambium. (b) In split stems that are approach-grafted, the cambia join with each other. (c) Grafting a stem with a severed cambium to one in which the cambium has not been damaged results in the severed cambium rejoining with itself, as in (a). (d) If the cambium is severed but the surface is covered with vaseline, then cambium does not form beneath the cut surface. (e) In a laterally displaced graft, the cambia link up in a way that preserves their radial polarity. (f) The gradient induction hypothesis assumes that there is a polar gradient across the region of the cambium and that new cambium forms only where cells having different positional specifications abut. (Warren Wilson & Warren Wilson 1984)

3.2.2). In any case, the speed at which florets were initiated, as fast as one every 10 min, and their spreading from one side of the plug, seems to preclude the operation of a diffusing inhibitor mechanism of the sort postulated for leaf initiation (see section 12.1.1). This development of the sunflower head seems more consistent with flower primordia having a promotory effect on the initiation of new primordia in the same way as different floral whorls seem to promote the formation of succeeding whorls rather than to inhibit them (see Ch. 4, section 4.4.2).

These examples of regeneration from higher plants show that small pieces of an organ, such as a shoot apex, can reconstitute themselves but the mechanism by which they do so is obscure. The reconstitution of the original shape would be consistent with the polar coordinate model of positional information, but it is not clear how this would result in the formation of a new structure, such as the procambium, that is not present in the original organ. The existence of polarity along plant axes, perhaps dependent on auxin flux or concentration, could perhaps correlate with the radial coordinates in the polar coordinate model (Fig. 12.13). In plants, there is also evidence for polarity in individual cells themselves and it seems to depend, as in the *Fucus* zygote or in the fern spore, or in the higher plant cells polarized for auxin transport, on the properties and structure of the plasma membrane. Each cell may become polarized longitudinally and radially as it forms. But how does a cell or group of cells then recognize that other cells are missing, and what is the signal to grow? Wound hormones released by the damaged cells or their walls may be important here in stimulating cell division and growth in the remaining cells. There is then the further question: how do the newly growing and dividing cells 'know' when the appropriate shape and cell mass has been reconstituted?

Size regulation has been modelled by supposing that some of the cells produce a substance A, which diffuses and is degraded throughout the whole organ. If part of the organ is removed, the concentration of A will tend to fall if producer cells have been removed or rise if non-producer cells are missing. The further assumption is that cell growth and multiplication are sensitive to the concentration of A and that when the concentration of A deviates from the predetermined concentration, then cell growth and division are stimulated until this concentration is achieved once more. This model can be extended to account for the regeneration of roots or shoots, if these are removed completely, and for the regulation of plant form. Since shoots are the main site of auxin production and roots the main site of cytokinin production, these substances could be involved in regulating the growth of other parts of the plant. Note that this model does not require that particular substances are made exclusively in certain parts of the plant – only that some parts are normally the principal sites of synthesis. So far, experiments have not been carried out specifically to test these hypotheses.

12.4.2   Regenerating cambia[8]

When part of a stem is cut away so that the vascular ring is interrupted, new cambium can be formed to link the severed tissues and restore the vascular ring (Fig. 12.14a). If two such damaged stems are approach grafted (Fig. 12.14b), then the ends of the cambia link up. When a stem with a severed vascular ring is grafted to one that has just had the outermost tissues removed (Fig. 12.14c), cambium forms to rejoin the severed ring as though the surface had been free as in Figure 12.14a. However, if the free surface is covered with vaseline or lanolin (Fig. 12.14d) then the ends of the cambium turn outwards but do not join up with each other and cambial continuity is not restored. When two half-stems are grafted so that the ends of the cambia that are opposed have opposite polarity, they do not link up directly but only by regeneration of new cambium so that the polarity is preserved (Fig. 12.14e).

These observations have been interpreted in terms of a gradient induction hypothesis, which states that morphogens are produced by the differentiated vascular tissues, creating a radial polarity (Fig. 12.14f), and that these morphogens diffuse into the adjacent cells and determine the position and polarity of the cambium that regenerates. Covering the cut surface with a layer impermeable to air, in some way prevents the setting up or persistence of the morphogen gradient so that cambium does not develop (Fig. 12.14d). The morphogens remain hypothetical, although auxin has been suggested as a candidate for one of them. The regeneration of cambia is one aspect of the regeneration of plant organs and, ideally, it should be explainable by a more general theory, such as the polar coordinate model, in which the cambium regeneration would occur because of an interruption of the polar coordinates and the polarity of the cambium would be restored in accordance with the radial coordinates (Fig. 12.13).

12.5   FORMATION OF PATTERN *DE NOVO* AND ITS PHYSICOCHEMICAL BASIS

Most development in plants is from cells and tissues that already show pattern, order, or polarity, and the new pattern can be traced to the old. But pattern has to originate some time. It may be at the formation of the cotyledons in the embryo or on production of the first leaves on a newly arisen, isolated shoot meristem in a tissue culture. Stomatal and hair patterns may sometimes arise by pattern formation in a previously uniform space. The first successful attempt to provide a plausible model to explain in molecular terms how a space might become patterned was made by A. M. Turing. He showed that stable patterns could be produced starting from a system in which the reactants were uniformly distributed. Two morphogens, X and Y, having different diffusion coefficients, interact in the presence of a catalyst, C, which is essentially therefore an evocator; Y is converted to X and the rate of degradation of both X and Y is dependent on the other substance. Starting with a uniform field of the reactants, small random fluctuations in this field would produce instability in the system, which would lead eventually to a stable pattern being set up. Patterns generated could include dapplings and standing waves. A standing wave on an annulus could specify the points at which morphogenetic events could occur, such as the formation of a whorl of leaves round a shoot apex. This type of hypothesis for pattern formation has been called a diffusion-reaction theory.

These ideas were the basis for the extensive and impressive work of H. Meinhardt, who has shown that patterns could be formed by the interaction of two processes: the autocatalytic production of an activator;

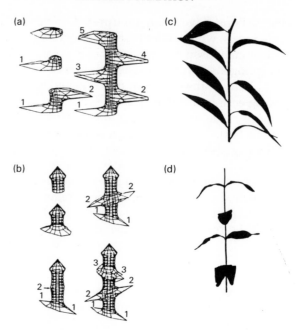

Figure 12.15 Models of leaf arrangement. Simulations of leaf formation according to Meinhardt's diffusion-reaction model. Stages in the formation of (a) a distichous and (b) a decussate leaf arrangement and examples of plants showing (c) distichous and (d) decussate leaf arrangement. (Meinhardt 1984)

and a process that acts antagonistically to the autocatalysis, either by the production of an inhibitor or by depletion of the substrate necessary for the autocatalytic reaction. Meinhardt's model systems can, with amazing accuracy, account for the development of polarity, the patterning of a uniform space, periodic patterns such as leaf formation (Fig. 12.15), branching patterns, the formation of complex patterns by a growing edge (as on mollusc shells), size regulation, and many other features of plant and animal development. An important point is that, counter-intuitively, the activator and inhibitor fields coincide and superimpose on each other. 'Common sense' would lead one to suppose that inhibitor maxima would be between the activator maxima, but this is not so (Fig. 12.16).

Specific substances acting as activators or inhibitors have not yet been identified. The important features of such molecules, on which the specific pattern formation would depend, would be their physical and metabolic properties, and therefore we would expect genes to specify pattern through their specification of the molecular nature of the morphogens and the enzyme systems by which they are metabolized. As Lindenmayer has remarked, 'morphogens and not genes are the basic concepts used in explaining development'. The whole pattern could be

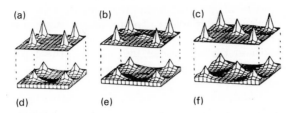

Figure 12.16 Formation of a new activator maximum during growth. As the surface with existing activator maxima expands (a, b) and the inhibitor concentration between them decreases (d, e), this allows a new activator maximum to develop (c) with its associated inhibitor maximum (f). Note that in this diffusion-reaction model, the activator and inhibitor maxima are superimposed and are not at complementary positions as might be supposed. (Meinhardt, 1984)

altered merely by the substitution of one molecule by a similar one, but with a different diffusion coefficient. Is this one explanation of the multiplicity of closely related plant growth substances? Are these candidates for morphogens in plant development? Plant growth substances spring to mind because they can and do control development in tissue cultures.

12.6   PATTERN FORMATION WITHIN ALGAL CELLS[9]

In multicellular organisms, pattern-forming processes will occur across cell boundaries. Only within single cells will patterns be formed within an uncompartmented system. Two examples of pattern formation have, therefore, been examined in single, large algal cells and compared with what would be expected according to a diffusion-reaction mechanism.

Box 12.1   *Acetabularia*[10]

Acetabularia is a remarkable unicellular alga that grows up to 20 cm long (Fig. 12.17). The single cell is held to the substratum by the rhizoid, formed from the basal part of the cell. The stalk grows, produces several whorls of hairs (which eventually drop off), and finally forms a cap. The nucleus, which is located in the rhizoid, divides there to form many secondary nuclei, which migrate up into the cap where they further divide to form the gametes, which are eventually released into the surrounding water. The morphology of the cap is distinctive for each species. Since the nucleus is in the rhizoid, the cell can be enucleated by cutting off the rhizoid. The stalk can continue to grow and develop without its nucleus and will even form a cap. When a rhizoid (containing its nucleus) from one species is grafted onto the stalk of another species, or the isolated nucleus from one species is transplanted into the enucleated stalk of another, the cap that forms is always characteristic of the species supplying the nucleus. The nucleus, therefore, controls cap morphology. The formation of a cap by an enucleated stalk implies that the nuclear products that ultimately determine cap shape persist in the absence of the nucleus. These products are long-lived mRNAs. The developmental sequence of changes in the synthesis of enzymes normally involved in DNA synthesis in the cap also depends on the presence of mRNAs but not on the actual presence of the nucleus. This means that the sequential expression of the genes coding for these enzymes is controlled indirectly through their products, the mRNAs, at the level of translation and not at transcription.

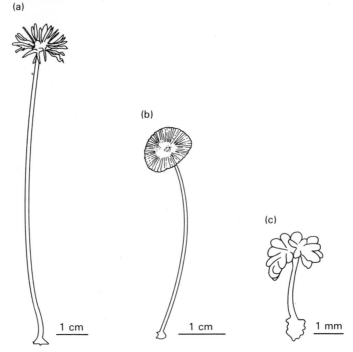

(a)

(b)

(c)

1 cm   1 cm   1 mm

Figure 12.17 Three species of *Acetabularia* differing in cap morphology: (a) *A. major*; (b) *A. mediterranea*; (c) *A. peniculus*. (After Berger *et al.* 1987)

Differentiation of different parts of the cell into rhizoid, stalk, and cap shows that this morphogenesis does not require compartmentation into cells. There must, therefore, be differences of some sort within different regions of the cytoplasm of the single cell. The cap-forming abilities of pieces of isolated stalk increase the nearer the tip of the plant from which the piece has been isolated. This suggests a gradient of increasing concentration of the necessary mRNAs towards the tip. The formation of a cap with rays implies that some parts of the periphery of the cell tip have grown out (to form the rays) but other parts (in between the rays) have not. How can this pattern arise as it does within a single cell? It has proved easier to tackle experimentally the problem of how a whorl of structures is formed by looking at the formation of the whorls of hairs on the stem.

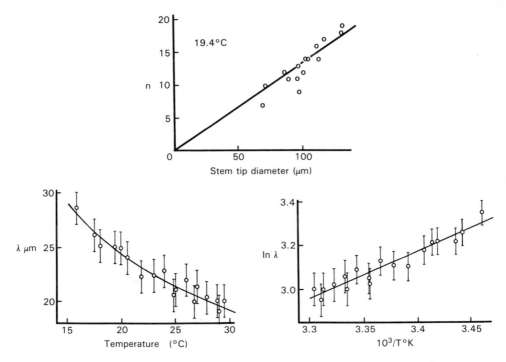

Figure 12.18 Formation of whorls of hairs in *Acetabularia*. (a) At a given temperature, the number of hairs per whorl (n) is proportional to stem tip diameter. (b) Spacing between hair initials (λ) decreases with increase in temperature. (c) The data of (b) replotted as an Arrhenius plot shows a linear relationship consistent with a mechanism dependent on chemical kinetics. (Harrison *et al.* 1981)

In the single-celled alga, *Acetabularia* (Box 12.1), successive whorls of hairs are formed on the stalk as it grows. Because the number of hairs in a whorl can be modified experimentally by altering the temperature, the pattern-forming process is amenable to experimental investigation. The number of hairs in a whorl at a given temperature is proportional to the diameter of the growing cell tip at the time of hair formation (Fig. 12.18a) and, therefore, the spacing between hair initials (λ) can be defined as $\lambda = \pi d/n$ (where d = tip diameter and n = number of hairs). Hair spacing decreases as temperature increases up to 30°C (Fig. 12.18b). When these data are replotted as an Arrhenius plot, a straight line is obtained (Fig. 12.18c). This is consistent with the spacing behaving like a chemical parameter and would, therefore, also be consistent with its being kinetically determined by varying concentrations of some chemical morphogen established in a wave pattern round the periphery of the growing cell tip. Cap formation seems to be similar and, since mRNAs specify the cap pattern, it may well be the proteins they code for or their products that are the diffusible substances setting up a pattern in which

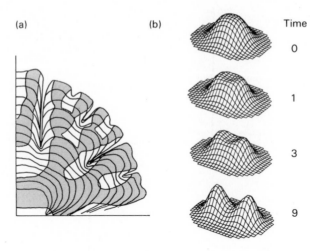

Figure 12.19 Pattern of wall formation in *Micrasterias*. (a) Radioactive labelling of the growing cell wall reveals a pattern of incorporation and synthesis that implies the formation at specific radii of new mutually inhibitory growing points at the edges of the old growing point. (b) This pattern is consistent with the pattern that would be expected according to diffusion-reaction theory and in which a broadening maximum would form two new maxima at its edges. (Lacalli & Harrison 1987)

there are regular maxima of activator molecules promoting local cell wall synthesis and so forming the cap rays.

Repeating patterns are also seen within single cells in the desmids, such as *Micrasterias*. The formation of the lobes of the enlarging semicell is by repeated bifurcations of the growing tips (see Ch. 6, Fig. 6.16). When cells were incubated with $[H^3]$ methylmethionine, this was incorporated by methylation into the polysaccharides of the growing cell wall. A pattern of growth was revealed that suggested that the lobe-producing centres widened and then branched when they reached a width that was specific for a given radius but that decreased with increasing radius (Fig. 12.19). This pattern is consistent with the operation of a diffusion-reaction system of the type proposed by Meinhardt.

Diatoms are much smaller and are remarkable for the detailed patterns of their silica shells. The pattern can develop in steps: in *Denticula*, the repeating pattern of ribs is first established and then the cross ribs gradually extend and grow from the centre line outwards (Fig. 12.20). Such patterns are the sort that can be simulated by a diffusion-reaction system, but nothing is yet known about how they arise in diatoms.

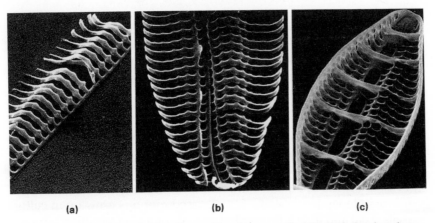

(a)   (b)   (c)

Figure 12.20 Pattern formation in a diatom. The siliceous wall of *Denticula* develops first as a series of ribs (a), which become joined by cross ribs (b) and eventually by septa (c). (Mann 1984; photographs kindly supplied by Dr D. G. Mann)

12.7   CELL–CELL COMMUNICATION[11]

For diffusion patterns to be set up in multicellular systems, there would have to be no effective barriers between the cells. There would need to be free passage of molecules across the plasmodesmata so that the symplast was continuous from cell to cell. The exclusion limit (upper size limit) for molecules to be able to pass freely through plasmodesmata is about 700 daltons, which would allow the passage of most metabolites and growth substances. In the shoot meristem, where some developmental patterns originate, small molecules diffuse freely whereas larger molecules do not and so substances such as plant growth substances could be candidates for activators and inhibitors in any diffusion-reaction processes involved in leaf positioning. Diffusion patterns could, presumably, be defined or altered by overall changes in plasmodesmatal properties or structure.

Cell-to-cell communication is important for development. When the cells of a fern protonema are **plasmolysed** to break symplasmic connections, each subsequently grows independently and reconstitutes a new protonema. The integrated growth of the original cells as a single unit is lost. Similarly, when higher plant embryo cells are isolated from each other, each has the potential to develop into a new embryo. Conversely, when tissues graft together, even when they are from such different genera as *Vicia* and *Helianthus*, plasmodesmata form between the grafted cells. Changes in Ca^{2+} ion concentration and distribution within cells may affect the functioning of plasmodesmata. There is some evidence that slight increases in intracellular Ca^{2+} may essentially cause blockage of the plasmodesmata. The possible role of Ca^{2+} in the integration of

279

growth has yet to be investigated. As a regulator of enzyme activities within cells, Ca^{2+} is undoubtedly involved, especially through its interaction with the protein calmoldulin, which, in turn, affects the activity of many enzymes. To understand the cellular mechanisms for the transduction of signals, from the environment and from neighbouring cells, is one of the major challenges of current research.

12.8   INTEGRATION OF DEVELOPMENT: PLANTS AS FAIL-SAFE SYSTEMS[12]

Individual organs of a plant can often continue to grow and develop to maturity independently of the rest. Excised root tips grow on in culture as roots. Leaf and flower primordia isolated soon after initiation can grow to maturity in culture, although they may not reach full size. Such excised organs can also often regenerate the missing parts of the plant, as an excised leaf may form roots or isolated roots may form shoot buds. In the intact plant, the growth and development of its potentially independent parts may be integrated by the supply of assimilates from shoot to root and of nutrients and metabolites from root to shoot. The many effects of growth substances when applied to plants (Table 12.2) provide strong evidence that they are key substances in controlling and integrating plant development. However, direct evidence for this is scanty except for a few systems, such as the control of stem length by gibberellins in peas and maize, as shown by the existence of gibberellin-deficient dwarf mutants. There are also many instances of changes in sensitivity of cells and organs to applied growth substances. This would be consistent with the growth substances themselves altering states of competence, but again there is as yet very little direct evidence that this is so in the intact plant (see Ch. 10).

The multiplicity of effects of growth substances on development is paralleled by the multiplicity of environmental signals that can elicit the same developmental response (Table 12.3). Many developmental processes can be triggered by many signals. Quite different signals can act as equally good triggers. Different sets of processes may also be able to act singly or in concert to produce the final result, as seems to be the case in leaf initiation and flowering (see Ch. 4, sections 4.4.1 & 4.4.2). It seems very unlikely that critical developmental processes such as these would be dependent on a single reaction sequence that could be fatally compromised by unfavourable environmental factors. The plant is at the mercy of its environment and it seems not at all unreasonable to suppose that it has back-up or fail-safe systems, in the same way that aircraft and spacecraft have two or three back-up computer systems, to avoid catastrophic failure.

It has been suggested by A. J. Trewavas that plant growth substances

Table 12.2 Promotive effects of growth substances applied to plants. The effects of the different types of growth substance may overlap or antagonize each other.

Auxins	Cell enlargement in excised stem and leaf tissue Cell division in cambium and (with cytokinin) in excised tissues and tissue cultures Initiation and development of vascular tissue Root initiation Growth of floral organs, especially sepals and stamens Femaleness in Cucurbits and fern gametophytes Apical dominance Delay of leaf senescence Delay of leaf and fruit abscission if applied distally Flowering in Bromeliads Epinastic responses
Gibberellins	Cell enlargement in excised stem and leaf tissue Stem elongation in rosette plants and some genetic dwarfs Leaf elongation in grasses and cereals Growth of floral organs Maleness in Cucurbits and fern gametophytes Growth of axillary buds and loss of apical dominance Fruit growth, especially seedless and stone fruits Bolting and flowering in long-day rosette plants and cold-requiring biennials Germination of seeds, especially light-sensitive seeds α-amylase synthesis in cereal grain aleurone cells
Cytokinins	Cell division Cell enlargement in excised stem and leaf tissue Shoot initiation in tissue cultures Growth of axillary buds and loss of apical dominance Delay of leaf senescence Germination of seeds, especially light-sensitive seeds
Abscisins	Stomatal closure Synthesis of seed storage proteins Dormancy of seeds and buds Inhibition of α-amylase synthesis in cereal grains
Ethylene	Fruit ripening Leaf and flower senescence Femaleness in Cucurbits Flowering in Bromeliads Loss of dormancy in buds and seeds Germination of seeds Epinastic responses

Table 12.3 Developmental responses that can be triggered in many plants by environmental and experimentally-produced signals. The same response may result from quite different signals, often in the same plant.

Plant response	Signals that can elicit the response
loss of seed and bud dormancy; germination	temperature; moisture; gibberellin; cytokinin; light
outgrowth of lateral buds	temperature; gibberellin; cytokinin; excision of terminal bud; mineral nutrients
flowering	day length; temperature; mineral nutrients; watering; irradiance; gibberellin; auxin; ethylene; sugar; orientation with respect to gravity
root initiation	moisture; auxin; ethylene
tropic curvatures	light; gravity; moisture
polarity of algal zygote outgrowth	various – see Table 6.1

represent just such a fail-safe system, which can act as a coordinator of growth and development in the absence of adequate assimilates or nutrients. In these circumstances, the growth substances would act as allocators of scarce resources, perhaps by controlling source and sink strengths. If plants do indeed have fail-safe systems, then the very act of experimenting on a plant may cause its growth and development to be diverted into alternative cellular pathways so that the intended effect of the experiment is nullified or modified, but of course without the experimenter being aware of this. Plants are developmental programs but perhaps with enough built-in redundancy that successful development is ensured even if parts of the program cannot be implemented because of unfavourable external or internal factors. Plants have necessarily had to adapt to tolerate their environment rather than create or choose their own environments as animals tend to do. It would, therefore, not be too surprising to find that, although using much the same cellular machinery as in animals, plant development differs from that in animals by having many distinctive features of its own.

12.9 SUMMARY

(1) The spacing of organs (leaves, roots, and branches) is strongly influenced by their near neighbours. Patterns are repeating and seem to depend only on local interactions within an autonomous pattern-generating system. Leaf arrangement can be accounted for in terms of diffusion-reaction mechanisms, in which the newly formed primordia represent maxima of activator and inhibitor in a

diffusion field. If a diffusion-reaction mechanism exists, it will interact with the biophysical pattern as expressed in the surface microstructure.

(2) Lateral roots form usually at specific points on the radius of the stele, which in a given species bear a constant relationship to the positions of the existing xylem and phloem poles. The longitudinal spacing of lateral roots could be the result of either a mutual inhibition mechanism or a cell lineage mechanism. Branching of shoots can be simulated by computer models, which assume only local information. Minimal assumptions necessary may often need to include interactions between units, such as an effect of apical dominance, which can be supported by experiment. The basis of branch angle may be structural, or a response to light or gravity, or both. Branching in the alga *Griffithsia* is a local program, apparently depending on interactions between neighbouring cells, which can be restarted by isolating single shoot cells from the filament.

(3) The patterning of vascular tissues is consistent with existing loci depleting their immediate area of metabolites so that new loci arise only when space is created by existing loci of differentiation becoming moved apart by growth.

(4) The patterns of stomata on the epidermis seem to arise not by the patterning of an undifferentiated space but by cell lineage mechanisms originating from polarized cell divisions in a field of cells all polarized along the same vector. The evidence that individual stomata inhibit the differentiation of other stomata is tenuous. Patterns of epidermal hairs also arise from polarized cell lineages. Differentiation of stomata and hairs can, though, be strongly influenced by the distribution of the underlying cells: stomata very often do not develop over veins, and hairs may develop only over intercellular spaces in the cortex.

(5) Shoot and root apices can regenerate themselves after parts have been removed surgically. Shoot apices can be divided into as many as six parts and each will regenerate a new apex, forming vascular tissue and leaves *de novo*. On regeneration of a new floral apex in sunflower, the new pattern of florets is consistent not with florets being mutually inhibitory but with their promoting the initiation of new florets. When cambia regenerate or form anew in damaged tissues, they join with existing cambia only in a way that preserves the radial polarity of the cambium. Diffusion-reaction models and the polar coordinate model can account for aspects of regeneration.

(6) The formation of pattern in unpatterned systems, especially within single algal cells, is consistent with diffusion-reaction models involving the interaction of an activator and an inhibitor of the initiation of units of the pattern. For movement through the sym-

plast, the upper molecular size (exclusion limit) is about 700 daltons so substances involved in diffusion-reaction mechanisms would be relatively small metabolites but could include the common growth substances. So far, no specific chemical substances have been identified as the proposed activators or inhibitors. The control of morphogenesis by the genes could be partly through the specification of active molecules with different diffusion coefficients.

(7) Many parts of plants are capable of independent growth. The integration of growth is probably through interchange of metabolites and growth substances in a relatively unspecific way. Growth substances may be involved in general integration and control of growth only in circumstances in which other controls have failed, although they may be crucial for pattern formation within the meristems. The multiplicity of effects of single growth substances, and of the different signals that can evoke a particular response, suggests that the plant has built into it many fail-safe systems and that redundancy of reaction mechanisms may be an essential adaptation of plants to their environment.

FURTHER READING

Barlow, P. W. & D. J. Carr (eds) 1984. *Positional controls in plant development*. Cambridge: Cambridge University Press.

Davies, P. J. (ed.) 1987. *Plant hormones and their role in plant growth and development*. Dordrecht: Martinus Nijhoff.

French, V., P. J. Bryant & S. V. Bryant 1976. Pattern regulation in epimorphic fields. *Science* **193**, 969–81. (Polar coordinate model of positional information)

Gunning, B. E. S. & A. W. Robards (eds) 1976. *Intercellular communication in plants: studies on plasmodesmata*. Berlin: Springer.

Hepler, P. K. & R. O. Wayne 1985. Calcium and plant development. *Annual Review of Plant Physiology* **36**, 397–439.

Lindenmayer, A. 1982. Developmental algorithms: lineage versus interactive control mechanisms. In *Developmental order: its origin and regulation*, S. Subtelny & P. B. Green (eds), 219–45. New York: Alan R. Liss.

Meinhardt, H. 1984. Models of pattern formation and their application to plant development. In *Positional controls in plant development*, P. W. Barlow & D. J. Carr (eds), 1–32. Cambridge: Cambridge University Press.

Sachs, T. 1984. Controls of cell patterns in plants. In *Pattern formation: a primer in developmental biology*, G. M. Malacinski (ed.), 367–91. New York: MacMillan.

NOTES

1 Barlow & Adam (1988). (Lateral root positioning)
2 Waaland (1984). (*Griffithsia*)
3 Cannell & Bowler (1978), Lindenmayer (1984). (Branching patterns)
4 Barlow & Rathfelder (1984), Feldman (1977), Torrey (1963). (Tissue patterns)

NOTES

5 Marx & Sachs (1977), Sachs (1974). (Stomatal patterns)
6 Barlow 1984, Cutter 1978. (Hair distribution)
7 Ball (1952), Feldman & Torrey (1976), Hernandez & Palmer (1988). (Regeneration of apical meristems)
8 Warren Wilson & Warren Wilson (1984). (Cambial regeneration)
9 Harrison et al. (1981), Lacalli & Harrison (1987), Mann (1984). (Pattern formation in algae)
10 Berger et al. (1987). (*Acetabularia*)
11 Goodwin & Erwee (1985), Kollman & Glockmann (1985). (Intercellular communication)
12 Trewavas (1986). (Plant growth substances)

285

Glossary

acropetal   Towards the apex. Basifugal (away from the base) would seem to be the same as acropetal, and acrofugal the same as **basipetal**; this distinction may be made if it is known how the process is influenced by the tip or base.

aleurone   The cell layer which surrounds the embryo and **endosperm** in cereal grains and which secretes amylase enzymes that break down the starch in the endosperm.

angiosperms   The seed plants in which the seed is enclosed in a fruit. In the gymnosperms (conifers and their relatives) the seeds are borne naked on scales, in cones, and not enclosed in a fruit. Angiosperms and gymnosperms together constitute the seed plants, or Spermaphyta.

antheridium   (see **bryophytes** and **pteridophytes**).

apical dome   The tip of the shoot apex distal to the youngest leaf **primordia** and often more or less hemispherical, although in many plants it can be almost flat.

apoplast   The part of the plant external to the plasma membrane, i.e. the cell walls, intercellular spaces, and dead cells such as **xylem**. Essentially the non-living part of the plant body. In a large tree much of the wood of the tree trunk, branches, and roots will be part of the apoplast (see **symplast**).

archegonium   (see **bryophytes** and **pteridophytes**).

axillary bud   A bud situated in the axil of a **leaf**, i.e. in the angle between the upper side of the leaf and the stem.

basipetal   Towards the base (see **acropetal**).

bract   A leaf-like structure immediately below a flower or inflorescence.

bryophytes   Liverworts and mosses; relatively simple plants. The reproductive structures are the antheridia (male) and archegonia (female), often borne on different plants. The gametes are motile and therefore fertilization depends on the plants being covered by a film of water in which the gametes move. The gametes are produced by the gametophytes (haploid). The zygote grows into the sporophyte (diploid), which appears as a stalk, surmounted by a sporangium, growing up from the leafy gametophyte. The sporophyte produces spores which are dispersed by wind. A spore germinates to form a protonema which is filamentous and consists of caulonema (main stem-like portion) and rhizoids (simple root-like structures). The protonema grows into the leafy gametophyte which is what we recognize as a moss or liverwort plant.

cambium   A cylinder of meristematic tissue which, by radial growth and anticlinal divisions (parallel to the surface), forms the secondary tissues in woody plants (most of the wood and the inner layers of the bark).

cell cycle   The progression of a cell from one mitosis to the next can be regarded as a cycle, the cell cycle. DNA synthesis usually occurs about the middle of interphase, midway between mitoses. The cell cycle can therefore be described as a succession of phases: G_1 – interphase preceding DNA synthesis; S – DNA synthesis; G_2 – interphase after DNA synthesis; and M – mitosis and cytokinesis (see also **polyploidy**).

cell division in plants   Mitosis and meiosis are essentially the same as in animals but the subsequent division of the cell, cytokinesis, differs. Almost all plant cells are surrounded by a cell wall and cytokinesis in such cells involves the separation of the two daughter cells by the formation of a new wall and not by cleavage of the cytoplasm. At telophase a cell plate is formed by the fusion of vesicles, containing wall material, which congregate across the cell in the position formerly occupied by the equator of the mitotic spindle. The cell plate forms the **middle lamella** on either side of which cell wall material is deposited. The cell plate grows and expands sideways, apparently pushing apart the **phragmoplast**, until it reaches the side walls where it fuses with them. Because the cell walls are fastened to each other no further movement is normally possible. Subsequent changes in cell size and shape can occur only by differential expansion of the cell wall and are necessarily shared by those adjacent cells which share walls.

coleoptile   A sheath enclosing the embryonic leaves in grass and cereal seedlings. Its growth is limited to several centimetres. The young leaves then break through its tip as they grow on. The coleoptile has been a classical object for the study of the effects of auxin on cell elongation.

cortex   The tissue system between the **epidermis** and the **stele**, often mainly of **parenchyma** but also often containing **sclerenchyma**. The innermost layer of the cortex is the **endodermis**, which surrounds the stele.

cotyledon   A seed leaf, specialized for storage of food material and formed in the embryo before the shoot apex is formed. Sometimes the cotyledons become transformed into the first leaves (e.g. tomato). In other plants they emerge from the soil but soon wither (e.g. runner beans); in others they remain below the soil (e.g. peas).

cuticle   A waxy secretion on the outside of the cell wall, especially on the outer (epidermal) surfaces of plants exposed to the air. It is relatively impermeable to water and gases.

desmids   Unicellular freshwater green algae in which the cell usually consists of two sculptured semicells, each the mirror image of the other, connected by a narrow isthmus in which lies the nucleus.

diatoms   Unicellular marine and freshwater yellow-brown algae with silicified cell walls.

dicotyledons   One of the two groups into which the **angiosperms** are divided and consisting of plants that have two cotyledons in the embryo. Dicotyledons characteristically have net-veined leaves, a ring of vascular bundles and a central pith in the stem, xylem in the root less than 7-arch (see **xylem poles**), cambium which gives secondary growth so that many dicotyledons are woody, and floral parts typically in fours or fives (see also **monocotyledons**).

distal   Nearer the tip (see **proximal**).

endodermis   The innermost layer of the cortex which surrounds the **stele** (the **vascular tissues** and pith) and which can act as a barrier to radial movement of solutes.

endosperm   The tissue formed in the developing seed by divisions of the nucleus, in the embryo sac, that results from fusion of one or more embryo sac nuclei with the haploid vegetative nucleus from the pollen tube. The resulting cells, usually triploid, form a nutritive tissue which in cereal grains is the main source of starch for the brewing industry. In many seeds (non-endospermous) the endosperm is digested and the products absorbed by the **cotyledons** which then become the main food store for the developing seedling (e.g. as in peas).

epicotyl   The stem above the **cotyledons** but below the first foliage leaves, most obvious in those seedlings in which the cotyledons remain below the soil.

epidermis   The outermost tissue of the plant, usually consisting of a single layer of cells (although sometimes it proliferates to give several layers) and covered by the **cuticle** except where there are stomatal apertures.

fibres   Elongated cells, typically pointed at both ends, with thick, lignified walls and having no protoplast when mature. **Sclerenchyma** is a tissue composed solely of fibres.

flowering, transition to   In many plants flowering is promoted by environmental signals, ensuring that individual members of a given species will tend to flower simultaneously, so enhancing the probability of successful cross fertilization. Promotion of flowering by **photoperiod** has been thought of as occurring in three major steps: induction of the leaves by the light stimulus to produce a signal; the action of this signal at the apex to commit the shoot apex to flowering, this process being evocation; and realization, or flower morphogenesis, the actual formation of the flowers. The action of photoperiod on the leaves but the stimulation of flowering at the shoot apex implies the production of some substance (hormone) or signal to transmit information from the leaves to the apex. The 'flowering hormone' ('florigen') has never been found. It is now thought that this idea is an oversimplification of what really happens, though what this is remains to be fully understood.

gametophyte   The gamete-producing (haploid or n) generation in plants (which in angiosperms is very much reduced) (see **bryophytes** and **pteridophytes**).

gymnosperms (see **angiosperms**).

homeotic mutants Mutants in which organs are replaced by other. organs not normally occurring in that position.

hypocotyl The stem below the **cotyledons** but above the root. Most obvious in seedlings in which the cotyledons emerge above the soil. The hypocotyl is where the transition between stem and root structure occurs.

inflorescence That part of the plant consisting of the flowers and the immediate branches or structures which bear them.

initial cells Those cells at the tip of the **meristem** which perpetuate the meristem and are the progenitors of all the other cells.

internode The stem between successive **nodes**. The internode is derived from the same modular unit of growth as the leaf above it, not the leaf below it as morphologists sometimes used to assume.

leaf An organ of limited growth, borne on the stem and having a bud in its axil (see **axillary bud**). A leaf is usually flattened dorso-ventrally into a blade (lamina) which is the main site of photosynthesis. The leaf stalk is the petiole and the blade may be subdivided into leaflets. Leaves originate as leaf primordia on the flanks of the shoot apex (see also **apical dome, primordia**).

light and plants The two main pigments in plants which are involved in transducing the action of light on developmental processes are phytochrome and cryptochrome. Phytochrome is red/far-red reversible and this unique property has been exploited to study it and its action. Cryptochrome is a pigment of which little is known but which is thought to be responsible for many of the effects of blue light (see **Light and Plant Growth** by J. W. Hart).

lignin The collective name for complex phenolic polymers found in cell walls, especially in **fibres** and **xylem** elements, and which make the cell wall rigid.

liverwort (see **bryophyte**).

meristem An organized tissue of apparently undifferentiated dividing cells, found at the apices of shoots and roots, in growing leaves and (as **cambium**) in stems and roots undergoing **secondary thickening**. Meristems may also be formed in callus and at the surface of wounds. Meristems are the source of new cells in the growing plant.

mesophyll The tissue of loosely packed chloroplast-containing cells making up the middle layer of the leaf and being the main site of photosynthesis.

microfibrils, cellulose The cellulose chains in cell walls are aggregated into bundles, or microfibrils, which intermesh with each other. Those parts of the microfibrils in which the cellulose chains are most highly ordered confer crystalline properties on the microfibrils and allow their examination using polarized light.

micropyle   The aperture left by the incomplete meeting of the integuments, the outermost layers of the **ovule**, and therefore the port of entry of the pollen tube as it grows into the ovule and gains access to the egg for fertilization.

middle lamella   That part of the cell wall which joins together the faces of walls of adjoining cells. It consists primarily of pectic substances and has a high content of calcium, thought to be involved in providing calcium pectate linkages which contribute to the middle lamella's cohesiveness (see also **cell division**).

monocotyledons   One of the two groups into which the **angiosperms** are divided, and consisting of plants that have only one **cotyledon** in the embryo. Monocotyledons characteristically have parallel-veined leaves, numerous vascular bundles distributed evenly throughout the stem, xylem in the root polyarch (see **xylem poles**), no **cambium** and so no **secondary thickening** (so not many are trees), and floral parts typically in threes or sixes. Monocotyledons include the grasses, cereals, bananas and palms (see also **dicotyledons**).

node   The point on a stem at which a leaf (or pair or whorl of leaves) is attached (see also **internode**).

nucellus   The tissue which is next to and encloses the embryo sac, the latter containing the egg and associated cells. The nucellus is itself enclosed by the integuments of the ovule (see Fig. 1.1).

ovary, ovule   The ovary is the female reproductive organ in the flower and consists of one or more carpels (free or fused together) each of which contains one or more ovules, each ovule containing an embryo sac within which is an egg cell (see Fig. 1.1).

parenchyma   A type of cell that is unspecialized, often isodiametric, sometimes elongated, and constituting most of the **cortex** in the stem and root and the **pith** in the stem.

pericycle   The outermost cell layer of the **stele**, and internal to the **endodermis**. Lateral roots originate from the pericycle. Sometimes **primordia** formed by the pericycle can develop into shoot buds, although these are usually formed from epidermal and cortical cells.

petiole   (see **leaf**).

phloem   A tissue of the vascular system consisting of sieve tubes, companion cells, and usually **parenchyma** and **fibres** as well. Its main function is the transport of metabolites from centres of assimilation and synthesis to centres of growth or storage. The transporting cells are the sieve elements, joined into longitudinal sieve tubes by the sieve plates between the cells. The sieve plates are perforated by the sieve pores, through which the cytoplasm is continuous from one sieve element to the next. Each sieve tube has alongside it one or more companion cells, which function in the loading and unloading of the sieve tubes. When phloem is damaged, an almost instantaneous response is the polymerization of callose, a polysaccharide that plugs the sieve pores and stops leakage of material from the phloem. The cytoplasm of the mature sieve tube is unusual in that the nucleus has disintegrated and the vacuole has merged with the cytoplasm. The cytoplasm is probably arranged into longitudinal strands that traverse the sieve pores and are intimately involved in the transport mechanism. In palm trees and other arborescent **monocotyledons** without secondary thickening the sieve tubes, once formed, have to function without nuclei for the whole life of the plant.

photoperiod Essentially daylength. Plants respond to the length of the light period in relation to the length of the preceding or succeeding dark period. Both 'light on' and 'light off' can be signals to initiate plant responses. In the induction of flowering it is often the length of the dark period that seems to be the more important. The length of a single photoperiodic cycle is usually 24 h but can be experimentally manipulated to be whatever length desired.

phragmoplast Microtubules derived from the mitotic spindle and which are displaced to the sides of the cell by the growing cell plate (see **cell division in plants**).

phyllotaxis The arrangement of leaves at the shoot apex and on the stem. These are not necessarily exactly the same, for as the stem matures the positions of the young leaves relative to each other may alter because of differential growth of the stem tissues. The mature arrangement also reflects the organization of the vascular system, to which the leaves are connected.

phytochrome (see **light and plants**).

pith The central **parenchymatous** tissue of the **stele**, and characteristic of the stem of **dicotyledons**.

pits Thin regions of the cell wall, usually containing numerous **plasmodesmata**.

plant growth substances (plant hormones) There are five classes of substances which have profound effects in modifying plant growth when they are applied experimentally at low concentrations. These are auxins, gibberellins, cytokinins, abscisins, and ethylene. They are all natural and ubiquitous components of plants. Originally thought of as hormones, being produced at one locus then moving to act at another, this is now believed to be an oversimplification and so the more neutral term 'growth substances' is preferred. The role of growth substances in the undisturbed plant is very unclear. The belief is that they act in the same ways, as demonstrated by external applications.

plasmodesmata Narrow cytoplasmic strands of complex structure through the cell wall which connect the cytoplasms of adjacent cells, so providing the continuity of cytoplasms which constitutes the **symplast**. Plasmodesmata do not normally seem to allow the passage of molecules larger than about 1000 daltons.

plasmolysis The withdrawal of the cytoplasm away from the cell wall due to loss of water from the vacuole. This results when the cell's

water potential is reduced below the point at which any degree of **turgor** can be maintained. Severe plasmolysis ruptures the **plasmodesmata**.

plastochron The time interval between the initiation of successive leaf **primordia**, or pairs or whorls of primordia at the shoot apex, and hence a measure of developmental time. More generally it is the interval between successive similar developmental events at the shoot apex.

plumule The embryonic shoot.

polyploidy More than two chromosome sets per cell. The normal 2 sets per cell is designated as $2n$, a gamete containing one set of n chromosomes. (Many plants are natural polyploids, in which the gametic set consists of a multiple of some basic number of chromosomes, x, so that in a tetraploid $2n = 4x$.) Polyploid cells can reach as high as $64n$ in the developing metaxylem. The n designation is not to be confused with C, which indicates the DNA level, the C amount of DNA being that in a gamete. A $2n$ nucleus may have either the $2C$, or $4C$ amount of DNA depending on whether it is in the G_1 or G_2 phase (before or after DNA synthesis respectively) in the cell cycle. An $8C$ DNA amount indicates that the nucleus is polyploid, but without further information it cannot be decided whether it is, for example, a tetraploid nucleus in the G_2 phase of the cell cycle or an octoploid nucleus in the G_1 phase.

primordia The early stage in the development of a root, shoot, or organ when it is just a small protuberance.

primordiomorph A group of cells resembling a **primordium** in size and position but in which there has been no corresponding increase in cell number.

procambium A tissue which differentiates in the apical and leaf **meristems** and which itself differentiates into the **xylem**, **cambium**, and **phloem** of the vascular system. The old term 'provascular strand' is a more accurate description than the modern term 'procambial strand'.

proembryo The early embryo before organs can begin to be distinguished.

promeristem Those cells in the meristem that are the immediate derivatives of the **initial cells**, are selfperpetuating, and maintain the **meristem** as other cells are displaced away and differentiate.

prothallus, protonema (see **bryophytes** and **pteridophytes**).

proximal Nearer the base (see **distal**).

pteridophytes Ferns, horsetails (*Equisetum*), clubmosses, and related plants, e.g. *Isoetes*, *Selaginella*. As in **bryophytes**, the reproductive structures are antheridia (male) and archegonia (female) and the gametes are motile. However, in the pteridophytes the gametophyte (haploid) which bears these reproductive structures and which produces the gametes, is small (up to about 1 cm across) and inconspicuous; when a thallus, and not filamentous, it is called the prothallus. It is often green and heart-shaped. Because the gametes require a film of moisture in which to move, fertilization in the prothalli occurs only in moist environments. The zygote grows into the sporophyte (diploid), which grows into the large, leafy sporophyte which we recognize as a typical fern plant. The spores, produced usually in millions by the sporophyte, are dispersed by wind. A spore germinates to form a protonema which is at first filamentous but in most species soon becomes two-dimensional to form the prothallus which is anchored to the substratum by **rhizoids**.

radicle   The embryonic root.

rays   Radial sheets of cells in the secondary **xylem** and **phloem**, which originate from the **cambium** and probably allow radial transfer of nutrients and assimilates to the cambium, which is sandwiched between the developing xylem and phloem that it has produced.

rhizoid   A filament of one or more cells apparently serving a root-like function in the lower (non-angiospermous) plants.

roots, adventitious   Roots growing not from other roots but from the stem or a leaf petiole or some other organ.

roots, lateral   Branch roots which originate in the **pericycle** a few millimetres or centimetres behind the parent root tip. Lateral roots themselves may bear further laterals and so on. Some grass plants have been estimated to have several kilometres of root.

sclerenchyma   (see **fibres**).

scutellum   The shield-like organ of the grass and cereal embryo that lies against the **endosperm** and acts as an absorptive organ for the embryo.

secondary thickening   The radial growth as a result of cambial activity which gives the growth in girth of woody plants (see **cambium**).

sieve elements, plates, pores, and tubes   (see **phloem**).

solenostele   The cylinder of vascular tissue in some ferns which is interrupted only by the small gaps immediately above the insertion of each leaf.

sporophyte   The spore-producing (diploid or $2n$) generation of all plants (see also **bryophytes** and **pteridophytes**).

stele   The innermost tissue system of the plant, surrounded by the **cortex** and **epidermis**. The outermost layer of the stele is the **pericycle** (from which lateral roots originate), which surrounds the vascular tissues, which in turn surround the central **pith** (which is characteristic of the stem but is present usually only in young roots).

stomata   (singular: *stoma*; plural can also be *stomates*) Each stoma is a pore in the **epidermis**, bounded by a pair of guard cells, which in turn are attached to subsidiary cells of the epidermis. The size of the pore is regulated by the guard cells, which change in shape to open the pore when they are turgid and to close it when they become less turgid. Their function is the regulation of gas exchange between the plant and the atmosphere.

symplast   The continuum of protoplasts in adjacent cells, the cytoplasms of which are linked by **plasmodesmata** and bounded by a common plasma membrane. Essentially the living part of the plant (see **apoplast**).

thallus   The plant body, sometimes amorphous, of the Algae and **bryophytes**, and of the gametophytic phase of most **pteridophytes**.

thin cell layer explants   Slices of stem **epidermis** plus a few subepidermal cell layers taken from flowering stalks of tobacco can be made to differentiate roots, leafy shoot buds, or flower buds, and eventually whole plants, according to the composition of the media on which they are cultured.

tip growth   Growth in which the maximum growth rate is at the extreme tip of the organ, the growth rate diminishing with distance from the tip. Typical of fungal hyphae, filamentous growth in lower plants, pollen tubes, and root hairs.

tissue   A collection of one or several cell types that are usually found together and are specialized for a particular function, e.g. **xylem** and **phloem**, specialized for transport. The three major divisions of the plant body – **epidermis**, **cortex**, and **stele** – may each be regarded as a tissue system. The stele is usually the most complex, consisting as it does of the vascular tissues in addition to **pith** and **pericycle**.

totipotency   The capacity of individual cells, immature or mature, to grow and divide to produce a new, whole organism.

tracheids   Individual **xylem** cells (xylem elements), usually very elongated, with characteristic pitting and wall thickenings. **Tracheary elements** are cells identifiable as xylem but often shorter or derived from **parenchymatous** cells.

transfer cells   Cells with very convoluted infoldings of parts of the cell wall and a correspondingly large surface area of plasma membrane. This provides a large surface for the passage of solute molecules into the cell. Transfer cells are found e.g. in **nodes**, and next to **phloem**, where there may be large fluxes of metabolites and solutes.

tunica, corpus   In the shoot apex of **angiosperms** it is usually possible to distinguish two layers by virtue of the planes of cell division in them. In the outermost layer, the tunica, consisting of the epidermis and often one or two layers of subepidermal cells, the new cell walls are all at right angles to the surface (anticlinal) and so these layers grow only in area. In the inner layers, the corpus, the plane of cell division may be at any angle, including being parallel to the surface (periclinal) so that growth in volume occurs. The first visible sign of leaf initiation is often the occurrence of periclinal divisions in tunica layers at the site of the new leaf primordium (see Fig. 2.4).

vascular tissues   A collection of tissues (**xylem** and **phloem**, often with **cambium** and **sclerenchyma**) specialized for transport, usually occurring as distinct vascular strands or vascular bundles which are usually arranged longitudinally in the plant and connect the roots, stems, leaves and other organs.

vernalization   The treatment of plants with low temperatures (usually

5–10°C is optimal), for several days or weeks, which promotes sub-sequent flowering.

vessels, xylem   These characteristic components of the xylem are made of many short xylem elements fused end to end with the end walls dissolved away so that they form very long tubes, sometimes several metres long. When mature they have lost all cellular contents and are non-living, and so are part of the **apoplast**. These are the main conduits for the upward movement of water and solutes from the roots to the shoot and other aerial parts of the plant. Xylem vessels are usually wider than other cells and so are prominent anatomical features of plants (see also **tracheids**).

water potential   The chemical potential of the water of the cells, measured in MegaPascals (units of pressure). A cell in equilibrium with water will be turgid and have a water potential of zero, which by definition is the water potential of pure water. When the cell becomes less turgid its water potential will become increasingly negative. Water movement in the plant is always down gradients of water potential.

xylem   A tissue of the vascular system and consisting of **vessels**, **tracheids**, **fibres**, and **parenchyma**. When mature, the vessels, tracheids, and fibres lack protoplasts and consist only of cell walls, which are usually lignified. The xylem is therefore a rigid tissue, especially in trees, where it forms the bulk of the plant, being the wood.

xylem poles   In a typical root the xylem appears star-shaped in cross section. In **dicotyledons** there are usually relatively few points to the star (about seven or usually fewer) whereas in **monocotyledons** there are many. Xylem with 2, 3, 4 ..... many poles is said to be di-, tri-, tetr- ..... poly-arch. The lateral roots often arise (in the **pericycle**) opposite the points or poles of the xylem (see Fig. 12.2) and so there are often as many longitudinal rows of lateral roots as there are poles to the xylem.

References

Allan, E. F. & A. J. Trewavas 1986. Tissue-dependent heterogeneity of cell growth in the root apex of *Pisum sativum*. *Botanical Gazette* **147**, 258–69.

Ball, E. 1952. Morphogenesis of shoots after isolation of the shoot apex of *Lupinus albus*. *American Journal of Botany* **39**, 167–91. (Shoot apex regeneration)

Ballade, P. 1970. Précisions nouvelles sur la caulogénèse apicale des racines axillaites du cresson (*Nasturtium officinale* R.Br). *Planta* **92**, 138–45. (Root–shoot transformation by cytokinin)

Barlow, P. W. 1969. Cell growth in the absence of division in a root meristem. *Planta* **88**, 215–23. (Inhibition of division but not growth with hydroxyurea)

Barlow, P. W. 1973. Mitotic cycles in root meristems. In *The cell cycle in development and differentiation*, M. Balls & F. S. Billet (eds), 113–65. Cambridge: Cambridge University Press.

Barlow, P. W. 1976. Towards an understanding of the behaviour of root meristems. *Journal of Theoretical Biology* **57**, 433–51. (Possible controls of cell growth and the significance of the quiescent centre)

Barlow, P. W. 1984. Positional controls in root development. In *Positional controls in plant development*, P. W. Barlow & D. J. Carr (eds), 281–318. Cambridge: Cambridge University Press. (Cell maturation as a function of position and time) (*Sinapis* roots)

Barlow, P. W. 1985. The nuclear endoreduplication cycle in metaxylem cells of primary roots of *Zea mays* L. *Annals of Botany* **55**, 445–57.

Barlow, P. W. 1987. Cellular packets, cell division and morphogenesis in the primary root meristem of *Zea mays* L. *New Phytologist* **105**, 27–56. (Analysis of cell division and growth in maize roots).

Barlow, P. W. & J. S. Adam 1988. The position and growth of lateral roots on cultured root axes of tomato, *Lycopersicon esculentum* (Solaneaceae). *Plant Systematics and Evolution* **158**, 141–54.

Barlow, P. W. & E. R. Hines 1982. Regeneration of the root cap of *Zea mays* L. and *Pisum sativum* L.: a study with the scanning electron microscope. *Annals of Botany* **49**, 521–9.

Barlow, P. W. & E. L. Rathfelder 1984. Correlations between the dimensions of different zones of grass root apices, and their implications for morphogenesis and differentiation in roots. *Annals of Botany* **53**, 249–60.

Basile, D. V. & M. R. Basile 1983. Desuppression of leaf primordia of *Plagiochila arctica* (Hepaticae) by ethylene antagonist. *Science* **220**, 1051–3. (Competence for leaf formation revealed by inhibiting wall extensin synthesis)

Bassel, A. R. 1985. Asymmetric cell division and differentiation: fern spore germination as a model. II. Ultrastructural studies. *Proceedings of the Royal Society Edinburgh* **86B**, 227–30. (Metal binding sites in fern spores)

Battey, N. H. & R. F. Lyndon 1988. Determination and differentiation of leaf and petal primordia in *Impatiens balsamina*. *Annals of Botany* **61**, 9–16.

Baulcombe, D. C., R. A. Martienssen, A. M. Huttley, R. F. Barker & C. M. Lazarus 1986. Hormonal and developmental control of gene expression in wheat. In *Differential gene expression and plant development*, C. J. Leaver, D. Boul-

296

ter & R. B. Flavell (eds). *Philosophical Transactions of the Royal Society London, Series B* **314**, 441–51.

Behrens, H. M., M. H. Weisenseel & A. Sievers 1982. Rapid changes in the pattern of electric current around the root tip of *Lepidium sativum* L. following gravistimulation. *Plant Physiology* **70**, 1079–83.

Benayoun, J., A. M. Catesson & Y. Czaninski 1981. A cytochemical study of differentiation and breakdown of vessel end walls. *Annals of Botany* **47**, 687–98.

Bennett, M. D. 1984. Towards a general model for spatial law and order in nuclear and karyotypic architecture. *Chromosomes Today* **8**, 190–202. (Organization of the interphase nucleus)

Berger, S., E. J. de Groot, G. Neuhaus & M. Schweiger 1987. Acetabularia: a giant single cell organism with valuable advantages for cell biology. *European Journal of Cell Biology* **44**, 349–70. (*Acetabularia*)

Blakely, L. M., R. M. Blakely, P. M. Colowit & D. S. Elliott 1988. Experimental studies on lateral root formation in radish seedling roots. II. Analysis of the dose–response to endogenous auxin. *Plant Physiology* **87**, 414–19.

Blakely, L. M., M. Durham, T. A. Evans & R. M. Blakely 1982. Experimental studies on lateral root formation in radish seedling roots. I. General methods, developmental stages and spontaneous formation of laterals. *Botanical Gazette* **143**, 341–52.

Bohdanowicz, J. 1987. *Alisma* embryogenesis: the development and ultrastructure of the suspensor. *Protoplasma* **137**, 71–83. (Basal cell as transfer cell).

Bonnett, H. T. & J. G. Torrey 1965. Chemical control of organ formation in root segments of *Convolvulus* cultured *in vitro*. *Plant Physiology* **40**, 1228–36. (Root and shoot induction by auxin and cytokinin)

Bowman, J. L., D. R. Smyth & E., M. Meyerowitz 1989. Genes directing flower development in *Arabidopsis*. *Plant Cell* **1**, 37–52.

Brandes, H. & H. Kende 1968. Studies on cytokinin-controlled bud formation in moss protonemata. *Plant Physiology* **43**, 827–37. (*Funaria*)

Brown, C. L. & K. Sax 1962. The influence of pressure on the differentiation of secondary tissues. *American Journal of Botany* **49**, 683–91.

Bruck, D. K. & D. J. Paolillo 1984. Replacement of leaf primordia with IAA in the induction of vascular differentiation in the stem of *Coleus*. *New Phytologist* **96**, 353–70.

Brulfert, J.L 1965. Etude expérimentale du développement végétatif et floral chez *Anagallis arvensis* L., ssp. *phoenicea* Scop. Formation de fleurs prolifères chez cette meme éspèce. *Révue Genérale de Botanique* **72**, 641–94. (Flower reversion)

Cannell, M. G. R. 1974. Production of branches and foliage by young trees of *Pinus contorta* and *Picea sitchensis*: provenance differences and their simulation. *Journal of Applied Ecology* **11**, 1091–115.

Cannell, M. G. R. 1976. Shoot apical growth and cataphyll initiation rates in provenances of *Pinus contorta* in Scotland. *Canadian Journal of Forestry Research* **6**, 539–56.

Cannell, M. G. R. & K. C. Bowler 1978. Spatial arrangement of lateral buds at the time that they form on leaders of *Picea* and *Larix*. *Canadian Journal of Forestry Research* **8**, 129–37. (Branching in conifers)

Caruso, J. L. & E. G. Cutter 1970. Morphogenetic aspects of a leafless mutant in tomato. II. Induction of a vascular cambium. *American Journal of Botany* **57**, 420–29. ('Reduced' mutant)

Chailakhyan, M.Kh. & V. N. Khryanin 1980. Hormonal regulation of sex expression in plants. In *Plant growth substances* 1979, F. Skoog (ed.), 331–44. Berlin: Springer. (Control of sexuality by growth substances)

Chailakhyan, M.Kh., N. P. Aksenova, T. N. Konstantinova & T. V. Bavrina 1975. The callus model of flowering. *Proceedings of the Royal Society London, Series B* **190**, 333–40.

Charles-Edwards, D. A., K. E. Cockshull, J. S. Horridge and J. H. M. Thornley 1979. A model of flowering in *Chrysanthemum*. *Annals of Botany* **44**, 557–66.

Christianson, M. L. & D. A. Warnick 1983. Competence and determination in the process of *in vitro* shoot organogenesis. *Developmental Biology* **95**, 288–93.

Christianson, M. L. & D. A. Warnick 1985. Temporal requirement for phytohormone balance in the control of organogenesis *in vitro*. *Developmental Biology* **112**, 494–7.

Cionini, P. G., A. Bennici, A. Alpi & F. D'Amato 1976. Suspensor, gibberellin and *in vitro* development of *Phaseolus coccineus* embryos. *Planta* **131**, 115–17.

Cleland, R. E. 1986. The role of hormones in wall loosening and plant growth. *Australian Journal of Plant Physiology* **13**, 93–103. (Auxin and wall extensibility)

Clowes, F. A. L. 1972. Regulation of mitosis in roots by their caps. *Nature New Biology* **235**, 143–4.

Clowes, F. A. L. 1981. The difference between open and closed meristems. *Annals of Botany* **48**, 761–7.

Coe, E. H. & M. G. Neuffer 1978. Corn embryo cell destinies. In *The clonal basis of development*, Society for Developmental Biology Symposium No. 36, S. Subtelny and I. M. Sussex (eds), 113–29. New York: Academic Press.

Cooke, T. J. & R. H. Racusen 1986. The role of electrical phenomena in tip growth, with special reference to the developmental plasticity of filamentous fern gametophytes. *Symposia of the Society for Experimental Biology* **40**, 307–28. (Ion currents in fern protonemata).

Cusick, F. 1956. Studies of floral morphogenesis. I. Median bisections of flower primordia in *Primula bulleyana* Forrest. *Transactions of the Royal Society Edinburgh* **63**, 153–66.

Cutter, E. G. 1954. Experimental induction of buds from fern leaf primordia. *Nature* **173**, 440–41.

Cutter, E. G. 1978. *Plant anatomy. Part I. Cells and tissues*, 2nd edn. London: Edward Arnold. (Cell types and their differentiation) (Hair formation and patterns, pp. 96–106)

Dean, C. & R. M. Leech 1982. Genome expression during normal leaf development. I. Cellular and chloroplast numbers and DNA, RNA, and protein levels in tissues of different ages within a seven-day-old wheat leaf. *Plant Physiology* **69**, 904–10. (Maturation of leaf cells)

DeMaggio, A. E. 1972. Induced vascular tissue differentiation in fern gametophytes. *Botanical Gazette* **133**, 311–17. (Induction of xylem in fern prothallus)

DeMaggio, A. E. 1982. Experimental embryology of pteridophytes. In *Experimental embryology of vascular plants*, B. M. Johri (ed.), 7–24. Berlin: Springer. (*Todea*, *Thelypteris*, and *Phlebodium*)

Dennin, K. A. & C. N. McDaniel 1985. Floral determination in axillary buds of *Nicotiana sylvestris*. *Developmental Biology* **112**, 377–82.

Deschamp, P. A. & T. J. Cooke 1984. Causal mechanisms of leaf dimorphism in the aquatic angiosperm *Callitriche heterophylla*. *American Journal of Botany* **71**, 319–29.

Dure, L. 1985. Embryogenesis and gene expression during seed formation. *Oxford Surveys of Plant Molecular and Cell Biology* **2**, 179–97.

Dyer, A. F. & M. A. L. King 1979. Cell division in fern protonemata. In *The experimental biology of ferns*, A. F. Dyer (ed.)., 307–54. London: Academic Press.

REFERENCES

Erickson, R. O. 1966. Relative elemental rates and anisotropy of growth in area: a computer program. *Journal of Experimental Botany* **17**, 390–403.

Erickson, R. O. & D. R. Goddard 1951. An analysis of root growth in cellular and biochemical terms. *Growth* **15**, 89–116. (Still the best kinetic analysis of root growth)

Esau, K. 1965. *Plant anatomy*, 2nd edn. New York: Wiley (Plant structure and anatomy, including embryogenesis briefly).

Evans, L. S. & J. Van't Hof 1974. Is the nuclear DNA content of mature root cells prescribed in the root meristem? *American Journal of Botany* **61**, 1104–11. (Proportions of mature root cells in G_1 and G_2)

Evert, R. F. & T. T. Kozlowski 1967. Effect of isolation of bark on cambial activity and development of xylem and phloem in trembling aspen. *American Journal of Botany* **54**, 1045–54.

Feldman, L. J. 1977. The generation and elaboration of primary vascular tissue patterns in roots of Zea. *Botanical Gazette* **138**, 393–401.

Feldman, L. J. & J. G. Torrey 1976. The isolation and culture *in vitro* of the quiescent center of Zea mays. *American Journal of Botany* **63**, 345–55.

Foard, D. E. 1971. The initial protrusion of a leaf primordium can form without concurrent periclinal cell divisions. *Canadian Journal of Botany* **49**, 1601–3.

Foard, D. E., A. H. Haber & T. N. Fishman 1965. Initiation of lateral root primordia without completion of mitosis and without cytokinesis in uniseriate pericycle. *American Journal of Botany* **52**, 580–90. (Primordiomorphs)

Fowler, M. W. & T. ApRees 1970. Carbohydrate oxidation during differentiation in roots of *Pisum sativum*. *Biochimica et Biophysica Acta* **201**, 33–44. (Respiratory pathways during cell maturation)

Francis, D. 1978. Regeneration of meristematic activity following decapitation of the root tip of *Vicia faba* L. *New Phytologist* **81**, 357–65. (Initiation and development of lateral roots as a function of time)

Francis, D. & R. F. Lyndon 1985. The control of the cell cycle in relation to floral induction. In *The cell division cycle in plants*, J. A. Bryant & D. Francis. (eds), Society for Experimental Biology, Seminar Series **26**, 199–215.

French, V., P. J. Bryant & S. V. Bryant 1976. Pattern regulation in epimorphic fields. *Science* **193**, 969–81. (Polar coordinate model of positional information)

Fry S. C. & E. Wangermann 1976. Polar transport of auxin through embryos. *New Phytologist* **77**, 313–17. (Auxin transport in *Phaseolus* and *Acer* embryos)

Fuchs, C. 1975. Ontogénese foliare et acquisition de la forme chez le *Tropaeolum peregrinum* L. *Annales des Sciences Naturelles Botanique* **16**, 321–90. (Growth rates and directions in leaf development)

Fukuda, H. & A. Komamine 1980. Direct evidence for cytodifferentiation to tracheary elements without intervening mitosis in a culture of single cells isolated from the mesophyll of *Zinnia elegans*. *Plant Physiology* **65**, 61–4.

Fukuda, H. & A. Komamine 1981. Relationship between tracheary element differentiation and the cell cycle in single cells isolated from the mesophyll of *Zinnia elegans*. *Physiologia Plantarum* **52**, 423–30.

Gahan, P. B. & L. M. Bellani 1984. Identification of shoot apical meristem cells committed to form vascular elements in *Pisum sativum* L. and *Vicia faba* L. *Annals of Botany* **54**, 837–41. (Cytochemical markers for early vascular differentiation)

Gersani, M. & T. Sachs 1984. Polarity reorientation in beans expressed by vascular differentiation and polar auxin transport. *Differentiation* **25**, 205–8. (Induction of transverse polarity of auxin transport)

Gertel, E. T. & P. B. Green 1977. Cell growth pattern and wall microfibrillar arrangement. Experiments with *Nitella*. *Plant Physiology* **60**, 247–54.

Gifford, E. M. 1983. Concept of apical cells in bryophytes and pteridophytes. *Annual Review of Plant Physiology* **34**, 419–40. (Review of apical cells)

Goldberg, R. B. 1989. Regulation of gene expression during plant embryogenesis. *Cell* **56**, 149–60.

Goldsmith, M. H. M. 1977. The polar transport of auxin. *Annual Review of Plant Physiology* **28**, 439–78. (Box 7.1)

Gonthier, R., A. Jacqmard & G. Bernier 1985. Occurrence of two cell subpopulations with different cell-cycle durations in the central and peripheral zones of the vegetative shoot apex of *Sinapis alba* L. *Planta* **165**, 288–91.

Gonthier, R., A. Jacqmard & G. Bernier 1987. Changes in cell cycle duration and growth fraction in the shoot meristem of *Sinapis* during floral induction. *Planta* **170**, 55–9.

Goodwin, P. B. & M. G. Erwee 1985. Intercellular transport studied by microinjection methods. In *Botanical microscopy 1985*, A. W. Robards (ed.), 335–58. Oxford: Oxford University Press. (Diffusion through the symplast)

Grayburn, W. S., P. B. Green & G. Steucek 1982. Bud induction with cytokinin. A local response to local application. *Plant Physiology* **69**, 682–6. (*Graptopetalum*)

Green, P. B. 1974. Morphogenesis of the cell and organ axis – biophysical models. *Brookhaven Symposia in Biology* **25**, 166–90.

Green, P. B. 1984. Shifts in plant cell axiality: histogenic influences on cellulose orientation in the succulent, *Graptopetalum*. *Developmental Biology* **103**, 18–27.

Green, P. B. 1985. Surface of the shoot apex: a reinforcement-field theory for phyllotaxis. *Journal of Cell Science Supplement* **2**, 181–201. (Surface structure and leaf initiation)

Green, P. B. 1988. A theory for inflorescence development and flower formation based on morphological and biophysical analysis in *Echeveria*. *Planta* **175**, 153–69. (Surface structure and flower formation)

Green, P. B. & K. E. Brooks 1978. Stem formation from a succulent leaf: its bearing on theories of axiation. *American Journal of Botany* **65**, 13–26.

Green, P. B. & R. S. Poethig 1982. Biophysics of the extension and initiation of plant organs. In *Developmental order: its origin and regulation*, S. Subtelny & P. B. Green (eds), 485–509. New York: Alan R. Liss.

Green, P. B., R. O. Erickson & P. A. Richmond 1970. On the physical basis of cell morphogenesis. *Annals of the New York Academy of Sciences* **175**, 721–31.

Grierson, D. 1986. Molecular biology of fruit ripening. *Oxford Surveys of Plant Molecular and Cell Biology* **3**, 363–83.

Gunning, B. E. S. 1978. Age-related and origin-related control of the numbers of plasmodesmata in cell walls of developing *Azolla* roots. *Planta* **143**, 181–90.

Gunning, B. E. S. 1982. The root of the water fern *Azolla*: cellular basis of development and multiple roles for cortical microtubules. In *Developmental order: its origin and regulation*, S. Subtelny & P. B. Green (eds), 379–421. New York: Alan R. Liss.

Gunning, B. E. S. & J. S. Pate 1974. Transfer cells. In *Dynamic aspects of plant ultrastructure*, A. W. Robards (ed.), 441–80. London: McGraw-Hill.

Gunning, B. E. S., J. E. Hughes & A. R. Hardham 1978. Formative and proliferative cell divisions, cell differentiation and developmental changes in the meristem of *Azolla* roots. *Planta* **143**, 121–44.

Haber, A. H. & D. E. Foard 1963. Nonessentiality of concurrent cell divisions for degree of polarization of leaf growth. II. Evidence from untreated plants and from chemically induced changes of the degree of polarization. *American Journal of Botany* **50**, 937–44.

Hackett, W. P., R. E. Cordero & C. Srinivasan 1987. Apical meristem characteristics and activity in relation to juvenility in *Hedera*. In *Manipulation of flowering*, J. G. Atherton (ed.), 93–9. London: Butterworth. (Characteristics of ivy)

Hardham, A. R. & M. E. McCully 1982. Reprogramming of cells following wounding in pea (*Pisum sativum* L.) roots. II. The effects of caffeine and colchicine on the development of new vascular elements. *Protoplasma* **112**, 152–66. (Inhibition of cell division)

Hardham, A. R., P. B. Green & J. M. Lang 1980. Reorganization of cortical microtubules and cellulose deposition during leaf formation in *Graptopetalum paraguayense*. *Planta* **149**, 181–95.

Harrison, L. G., J. Snell, R. Verdi, D. E. Vogt, G. D. Zeiss & B. R. Green 1981. Hair morphogenesis in *Acetabularia mediterranea*: temperature-dependent spacing and models of morphogen waves. *Protoplasma* **106**, 211–21. (Diffusion-reaction model)

Hejnowicz, Z. & P. Brodski 1960. The growth of cells as a function of time and their position in the root. *Acta Societatis Botanicorum Poloniae* **29**, 625–44. (Cell lengths as time markers)

Henry, Y. & M. W. Steer 1980. A re-examination of the induction of phloem transfer cell development in pea leaves (*Pisum sativum*). *Plant Cell and Environment* **3**, 377–80. (Induction of transfer cells)

Hernandez, L. F. & J. H. Palmer 1988. Regeneration of the sunflower capitulum after cylindrical wounding of the receptacle. *American Journal of Botany* **75**, 1253–61.

Heyes, J. K. & R. Brown 1965. Cytochemical changes in cell growth and differentiation in plants. *Encyclopedia of Plant Physiology*, Vol. XV/I, W. Ruhland (ed.), 189–212.

Heyes, J. K. & D. Vaughan 1967. The effects of 2-thiouracil on growth and metabolism in the root. I. Growth of excised root tissue. *Proceedings of the Royal Society London, Series B* **169**, 77–88. (Growth of isolated root segments)

Houck, D. F. & C. E. LaMotte 1977. Primary phloem regeneration without concomitant xylem regeneration: its hormone control in *Coleus*. *American Journal of Botany* **64**, 799–809. (Also cytokinin requirement in excised internodes)

Jackson, J. A. & R. F. Lyndon 1988. Cytokinin habituation in juvenile and flowering tobacco. *Journal of Plant Physiology* **132**, 575–9.

Jacobs, W. P. 1979. *Plant hormones and plant development*. Cambridge: Cambridge University Press. (An individualistic view, including vascular regeneration around wounds)

Jacobs, M. & S. F. Gilbert 1983. Basal localization of the presumptive auxin transport carrier in pea stem cells. *Science* **220**, 1297–3000. (Box 7.1)

Jaffe, L. F. 1966. Electrical currents through the developing *Fucus* egg. *Proceedings of the National Academy of Science, U.S.A.* **56**, 1102–9. (*Fucus* zygotes in tubes)

Jaffe, L. F. 1969. Localization in the developing *Fucus* egg and the general role of localizing currents. *Advances in Morphogenesis* **7**, 295–328.

Jaffe, L. F. & R. Nuccitelli 1974. An ultrasensitive vibrating probe for measuring steady extracellular currents. *Journal of Cell Biology* **63**, 614–28. (The vibrating probe electrode)

Jeffs, R. A. & D. H. Northcote 1967. The influence of indol-3-yl acetic acid and sugar on the pattern of induced differentiation in plant tissue culture. *Journal of Cell Science* **2**, 77–88. (Differentiation of vascular tissues at specific points on a diffusion gradient in bean callus)

REFERENCES

Jegla, D. E. & I. M. Sussex 1987. Clonal analysis of meristem development. In *Manipulation of flowering*, J. G. Atherton (ed.), 101–8. London: Butterworth.

Jensen, L. C. W. 1971. Experimental bisection of *Aquilegia* floral buds cultured *in vitro*. I. The effect on growth, primordia initiation, and apical regeneration. *Canadian Journal of Botany* **49**, 487–93. (Surgical experiments on developing flowers)

Jensen, W. A. 1976. The role of cell division in angiosperm embryology. In *Cell division in higher plants*, M. M. Yeoman (ed.), 391–405. London: Academic Press. (Cell division in cotton embryo)

Jensen, W. A. & M. Ashton 1960. Composition of the developing primary wall in onion root tip cells. I. Quantitative analysis. *Plant Physiology* **35**, 313–23.

Jeune, B. 1975. Croissance des feuilles aeriennes de *Myriophyllum brasiliense* Camb. *Adansonia* **15**, 257–71. (Growth rates and directions in leaf development)

Kallio, P. & J. Lehtonen 1981. Nuclear control of morphogenesis in *Micrasterias*. In *Cytomorphogenesis in plants*, O. Kiermayer (ed.), 191–228. Vienna: Springer.

Kiermayer, O. 1981. Cytoplasmic basis of morphogenesis in *Micrasterias*. In *Cytomorphogenesis in plants*, O. Kiermayer (ed.), 147–89. Vienna: Springer.

Kinet, J.-M., R. M. Sachs & G. Bernier 1985. *The physiology of flowering*, Vol. III. Boca Raton: CRC Press. (Growth substances and correlative growth in flowers)

Kinet, J.-M., M. Bodson, A. M. Alvinia & G. Bernier 1971. The inhibition of flowering in *Sinapis alba* after the arrival of the floral stimulus at the meristem. *Zeitschrift für Pflanzenphysiologie* **66**, 49–63.

King, R. W. & L. T. Evans 1969. Timing of evocation and development of flowers in *Pharbitis nil*. *Australian Journal of Biological Science* **22**, 559–72.

Kollman, R. & C. Glockmann 1985. Studies on graft union. I. Plasmodesmata between cells of plants belonging to different unrelated taxa. *Protoplasma* **124**, 224–35. (Cell–cell communication between *Vicia* and *Helianthus*)

Komaki, M. K., K. Okada, E. Nishino & Y. Shimura 1988. Isolation and characterization of novel mutants of *Arabidopsis thaliana* defective in flower development. *Development* **104**, 195–203.

Koshland, D. E., T. J. Mitchison & M. W. Kirschner 1988. Polewards chromosome movement driven by microtubule depolymerization *in vitro*. *Nature* **331**, 499–504.

Kotenko, J. L., J. H. Miller & A. I. Robinson 1987. The role of asymmetric cell division in pteridophyte cell differentiation. I. Localized metal accumulation and differentiation in *Vittaria* gemmae and *Onoclea* prothalli. *Protoplasma* **136**, 81–95.

Kropf, D. L., S. K. Berge & R. Quatrano 1989. Actin localization during *Fucus* embryogenesis. *Plant Cell* **1**, 191–200.

Kropf, D. L., B. Kloareg & R. S. Quatrano 1988. Cell wall is required for fixation of the embryonic axis in *Fucus* zygotes. *Science* **239**, 187–90.

Kulkarni, V. J. & W. W. Schwabe 1984. Differences in graft transmission of the floral stimulus in two species of *Kleinia*. *Journal of Experimental Botany* **35**, 422–30.

Kutschera, U., R. Bergfeld & P. Schopfer 1987. Cooperation of epidermis and inner tissues in auxin-mediated growth of maize coleoptiles. *Planta* **170**, 168–80.

Lacalli, T. C. & L. G. Harrison 1987. Turing's model and branching tip growth: relation of time and spatial scales in morphogenesis, with application to *Micrasterias*. *Canadian Journal of Botany* **65**, 1308–19. (Diffusion-reaction model)

Larson, P. R. 1983. Primary vascularization and the siting of primordia. In *The growth and functioning of leaves*, J. E. Dale & F. L. Milthorpe (eds), 25–51. Cambridge: Cambridge University Press.

Lindenmayer, A. 1982. Developmental algorithms: lineage versus interactive control mechanisms. In *Developmental order: its origin and regulation*, S. Subtelny & P. B.. Green (eds), 219–45. New York: Alan R. Liss.

Lindenmayer, A. 1984. Positional and temporal control mechanisms in inflorescence development. In *Positional controls in plant development*, P. W. Barlow & D. J. Carr (eds), 461–86. Cambridge: Cambridge University Press. (Branching models)

Lintilhac, P. M. 1974. Positional controls in meristem development: a caveat and an alternative. In *Positional controls in plant development*, P. W. Barlow & D. J. Carr (eds), 83–105. Cambridge: Cambridge University Press. (Effects of pressure on callus)

Lintilhac, P. M. & T. B. Vesecky 1980. Mechanical stress and cell wall orientation in plants. I. Photoelastic derivation of principal stresses. With a discussion of the concept of axillarity and the significance of the 'arcuate shell zone'. *American Journal of Botany* **67**, 1477–83.

Lloyd, C. W., L. Clayton, P. J. Dawson, J. H. Doonan, J. S. Hulme, I. N. Roberts & B. Wells 1985. The cytoskeleton underlying side walls and cross walls in plants: molecules and macromolecular assemblies. *Journal of Cell Science Supplement* **2**, 143–55.

Lyndon, R. F. 1973. The cell cycle in the shoot apex. In *The cell cycle in development and differentiation*, M. Balls & F. S. Billett (eds), 167–83. Cambridge: Cambridge University Press.

Lyndon, R. F. 1976. The shoot apex. In *Cell Division in higher plants*, M. M. Yeoman (ed.), 285–314. London: Academic Press.

Lyndon, R. F. 1978. Phyllotaxis and the initiation of primordia during flower development in *Silene*. *Annals of Botany* **42**, 1349–60.

Lyndon, R. F. 1979. The cellular basis of apical differentiation. In *Differentiation and the control of development in plants – potential for chemical modification*, E. C. George (ed.). *British Plant Growth Regulator Group Monograph* **3**, 57–73. (Cell maturation as a function of time rather than position)

Lyndon, R. F. 1983. The mechanism of leaf initiation. In *The growth and functioning of leaves*, J. E. Dale & F. L. Milthorpe (eds), 3–24. Cambridge: Cambridge University Press.

Lyndon, R. F. 1987. Initiation and growth of internodes and stem and flower frusta in *Silene coeli-rosa*. In *The manipulation of flowering*, J. Atherton (ed.), 301–14. London: Butterworth.

Lyndon, R. F. & N. H. Battey 1985. The growth of the shoot apical meristem during flower initiation. *Biologia Plantarum* **27**, 339–49.

Lyndon, R. F. & M. E. Cunninghame 1986. Control of shoot apical development via cell division. In *Plasticity in plants*, D. H. Jennings & A. J. Trewavas (eds). *Symposia of the Society for Experimental Biology* **40**, 233–55. (Leaf initiation and surface divisions)

Lyndon, R. F. & D. Francis 1984. The response of the shoot apex to light generated signals from the leaves. In *Light and the flowering process*, D. Vince-Prue, B. Thomas & K. E. Cockshull (eds), 171–89. London: Academic Press.

Lyndon, R. F. & E. S. Robertson 1976. The quantitative ultrastructure of the pea shoot apex in relation to leaf initiation. *Protoplasma* **87**, 387–402. (Changes in cell organelle numbers during early differentiation)

MacLeod, R. D. & D. Francis 1976. Cortical cell breakdown and lateral root primordium development in *Vicia faba* L. *Journal of Experimental Botany* **27**, 922–32.

Mann, D. G. 1984. An ontogenetic approach to diatom systematics. In *Proceedings 7th International Diatom Symposium*, D. G. Mann (ed.), 113–44. Koenigstein: O. Koeltz.

Marx, A. & T. Sachs 1977. The determination of stomata pattern and frequency in *Angallis*. *Botanical Gazette* **138**, 385–92.

Masuda, Y. & R. Yamamoto 1985. Cell-wall changes during auxin-induced cell extension. Mechanical properties and constituent polysaccharides of the cell wall. In *Biochemistry of plant cell walls*, C. T. Brett & J. R. Hillman (eds). *Society for Experimental Biology Seminar Series* **28**, 269–300. (Changes in wall chemistry during cell extension)

Meicenheimer, R. D. 1981. Changes in *Epilobium* phyllotaxy induced by N-1-naphthylphthalamic acid and a-4-chlorophenoxyisobutyric acid. *American Journal of Botany* **68**, 1139–54.

Meindl, U. 1982. Local accumulation of membrane-associated calcium according to cell pattern formation in *Micrasterias denticulata*, visualized by chlorotetracycline fluorescence. *Protoplasma* **110**, 143–6.

Meinhardt, H. 1984. Models of pattern formation and their application to plant development. In *Positional controls in plant development*, P. W. Barlow & D. J. Carr (eds), 1–32. Cambridge: Cambridge University Press.

Meinke, D. W. 1986. Embryo-lethal mutants and the study of plant embryo development. *Oxford Surveys of Plant Molecular and Cell Biology* **3**, 122–65. (*Arabidopsis*)

Meins, F. & A. N. Binns 1978. Epigenetic clonal variation in the requirement of plant cells for cytokinins. In *The clonal basis of development*, Symposia of the Society for Developmental Biology No. 36, S. Subtelny & I. M. Sussex (eds), 185–201. New York: Academic Press. (Habituation)

Meins, F. & A. N. Binns 1979. Cell determination in plant development. *BioScience* **29**, 221–5.

Meins, F. & J. Lutz 1979. Tissue-specific variation in the cytokinin habituation of cultured tobacco cells. *Differentiation* **15**, 1–6.

Miginiac, E. 1972. Cinetique d'action comparée des racines et de la kinétine sur le développement floral de bourgeons cotylédonaires chez le *Scrofularia arguta* Sol. *Physiologie Végétale* **10**, 627–36. (Inhibition of flower initiation by roots and cytokinins)

Miller, J. H. 1980. Orientation of the plane of cell division in fern gametophytes: the roles of cell shape and stress. *American Journal of Botany* **67**, 534–42. (Transition to 2-D growth)

Miller, J. H. 1985. Asymmetric cell division and differentiation; fern spore germination as a model. I. Physiological aspects. In *Biology of pteridophytes*, A. F. Dyer & C. N. Page (eds). *Proceedings of the Royal Society Edinburgh* **86B**, 213–26. (Unequal division in fern spores)

Miller, D. R. & J. R. Goodin 1976. Cellular growth rates of juvenile and adult *Hedera helix* L. *Plant Science Letters* **7**, 397–401. (Induction of mature to juvenile phase change in ivy callus by gibberellic acid)

Mineyuki, Y. & M. Furuya 1980. Effect of centrifugation on the development and timing of premitotic positioning of the nucleus in *Adiantum* protonemata. *Development Growth and Differentiation* **22**, 867–74. (Induction of branching by displacement of nucleus)

Mineyuki, Y. & M. Furuya 1986. Involvement of colchicine- sensitive cytoplasmic element in premitotic nuclear positioning of *Adiantum* protonemata. *Protoplasma* **130**, 83–90.

Mullins, M. G. 1980. Regulation of flowering in the grapevine. In *Plant growth substances, 1979*, F. Skoog (ed.), 323–30. Berlin: Springer. (*Vitis*)

REFERENCES

Naf, U. 1979. Antheridiogens and antheridial development. In *The experimental biology of ferns*, A. F. Dyer (ed.), 435–70. London: Academic Press. (Fern prothalli)

Navarette, M. H. & C. Bernabeu 1978. Soluble polypeptides from meristematic and mature cells of *Allium cepa* roots. *Planta* **142**, 147–51. (Changes in protein complement)

Northcote, D. H. 1963. Changes in the cell walls of plants during differentiation. *Symposia of the Society for Experimental Biology* **17**, 157–74. (Changes in chemical composition of xylem walls during differentiation)

Ormrod, J. & D. Francis 1986. Mean rate of DNA replication and replicon size in the shoot apex of *Silene coeli-rosa* L. during the initial 120 minutes of the first day of floral induction. *Protoplasma* **130**, 206–10.

Palevitz, B. A. 1981. The structure and development of stomatal cells. In *Stomatal physiology*, P. G. Jarvis & T. A. Mansfield (eds). Society for Experimental Biology Seminar Series 8, 1–23. Cambridge: Cambridge University Press.

Palevitz. B. A. & P. K. Hepler 1974. The control of the plane of division during stomatal differentiation in *Allium*. I. Spindle reorientation. *Chromosoma* **46**, 297–326.

Philipson, J. J. & M. P. Coutts 1980. Effects of growth hormone application on the secondary growth of roots and stems in *Picea sitchensis* (Bong.) Carr. *Annals of Botany* **46**, 747–55. (Production of rays by cytokinin)

Phillips, R. 1981. Direct differentiation of tracheary elements in cultured explants of gamma-irradiated tubers of *Helianthus tuberosus*. *Planta* **153**, 262–6. (Inhibition of cell division)

Pickett-Heaps, J. D. 1969. Preprophase microtubules and stomatal differentiation in *Commelina cyanea*. *Australian Journal of Biological Science* **22**, 375–91. (Cell divisions to form stomata)

Pilet, P.-E. & P. W. Barlow 1987. The role of abscisic acid in root growth and gravireaction: a critical review. *Plant Growth Regulation* **6**, 217–65.

Poethig, S. 1987. Clonal analysis of cell lineage patterns in plant development. *American Journal of Botany* **74**, 581–94. (Review of clonal analysis)

Preston, R. D. 1974. *The physical biology of plant cell walls*. London: Chapman & Hall. (Wall structure and methods of investigation including use of polarized light)

Quatrano, R. S., L. R. Griffing, V. Huber-Walchli & R. S. Doubet 1985. Cytological and biochemical requirements for the establishment of a polar cell. *Journal of Cell Science Supplement* **2**, 129–41.

Raghavan, V. 1976. *Experimental embryogenesis in vascular plants*. London: Academic Press. (Irradiation of embryos; *Cuscuta*)

Raghavan, V. & P. S. Srivastava 1982. Embryo culture. In *Experimental embryology of vascular plants*. B. M. Johri (ed.), 195–230. Berlin: Springer. (Effects of osmotic concentration and growth hormones)

Reiss, H. D. & W. Herth 1978. Visualization of the Ca^{2+} gradient in growing pollen tubes of *Lilium longiflorum* with chlorotetracycline fluorescence. *Protoplasma* **97**, 373–7.

Reiss, H. D. & W. Herth 1979. Calcium gradients in tip growing plant cells visualized by chlorotetracycline fluorescence. *Planta* **146**, 615–21.

Rier, J. P. & D. T. Beslow 1967. Sucrose concentration and the differentiation of xylem cells. *Botanical Gazette* **128**, 73–7. (*Parthenocissus* callus)

Roberts, J., J. Burgess, I. Roberts & P. Linstead 1985. Microtubule rearrangement during plant cell growth and development: an immunofluorescence study. In *Botanical microscopy 1985*, A. W. Robards (ed.), 263–83. Oxford: Oxford University Press.

Robinson, K. R. & L. F. Jaffe 1975. Polarizing fucoid eggs drive a calcium current through themselves. *Science* **187**, 70–72.

Robinson, A. I., J. H. Miller, R. Helfrich & M. Downing 1984. Metal-binding sites in germinating fern spores (*Onoclea sensibilis*). *Protoplasma* **120**, 1–11.

Rodriquez, D., J. Dommes & D. H. Northcote 1987. Effect of abscisic and gibberellic acids on malate synthase transcripts in germinating castor bean seeds. *Plant Molecular Biology* **9**, 227–35.

Rubery, P. H. 1987. Auxin transport. In *Plant hormones and their role in growth and development*, P. J. Davies (ed.), 341–62. Dordrecht: Martin Nijhoff.

Sachs, T. 1968. On the determination of the pattern of vascular tissue in peas. *Annals of Botany* **32**, 781–90.

Sachs, T. 1969. Regeneration experiments on the determination of the form of leaves. *Israel Journal of Botany* **18**, 21–30.

Sachs, T. 1969. Polarity and the induction of organized vascular tissues. *Annals of Botany* **33**, 263–75.

Sachs, T. 1974. The developmental origin of stomata pattern in *Crinum*. *Botanical Gazette* **135**, 314–18.

Sachs, T. 1981. The control of patterned differentiation of vascular tissues. *Advances in Botanical Research* **9**, 151–262. (Summary and synthesis of work on induction of vascular strands)

Sachs, T. 1984. Axiality and polarity in plants. In *Positional controls in plant development*, P. W. Barlow & D. J. Carr (eds), 193–224. Cambridge: Cambridge University Press.

Sakaguchi, S., T. Hogetsu & N. Hara 1988. Arrangements of cortical microtubules in the shoot apex of *Vinca major* L. Observations by immunofluorescence microscopy. *Planta* **175**, 403–11.

Savidge, R. A. & P. F. Wareing 1981. Plant-growth regulators and the differentiation of vascular elements. In *Xylem cell development*, J. R. Barnett (ed.), 192–235. Tunbridge Wells, Castle House Publications. (Control of differentiation of secondary vascular tissues)

Sawhney, V. K. & R. I. Greyson 1979. Interpretations of determination and canalization of stamen development in a tomato mutant. *Canadian Journal of Botany* **57**, 2471–7.

Schwabe, W. W. 1971. Chemical modification of phyllotaxis and its implications. *Symposia of the Society for Experimental Biology* **25**, 301–22.

Schwabe, W. W. & A. H. Al-Doori 1973. Analysis of a juvenile-like condition affecting flowering in the black currant (*Ribes nigrum*). *Journal of Experimental Botany* **24**, 969–81.

Selman, G. 1966. Experimental evidence for nuclear control of differentiation in *Micrasterias*. *Journal of Embryology and Experimental Morphology* **16**, 469–85. (Nuclear control of *Micrasterias* wall growth)

Shabde, M. & T. Murashige 1977. Hormonal requirements of excised *Dianthus caryophyllus* L. shoot apical meristems *in vitro*. *American Journal of Botany* **64**, 443–8.

Siebers, A. M. 1971. Initiation of radial polarity in the interfascicular cambium of *Ricinus communis* L. *Acta Botanica Neerlandica* **20**, 211–20. (Experiments on orientation of cambium)

Sievers, A. & E. Schnepf 1981. Morphogenesis and polarity of tubular cells with tip growth. In *Cytomorphogenesis in plants*. O. Kiermayer (ed.), 265–99. Vienna: Springer.

Sinnott, E. W. 1944. Cell polarity and the development of form in cucurbit fruits. *American Journal of Botany* **31**, 388–91.

Sinnott, E. W. & R. Bloch 1945. The cytoplasmic basis of intercellular patterns in vascular differentiation. *American Journal of Botany* **32**, 151–6.

Smart, C. C. & N. Amrhein 1985. The influence of lignification on the development of vascular tissue in *Vigna radiata* L. *Protoplasma* **124**, 87–95. (Inhibition of PAL and lignification in xylogenesis)

Snow, M. & R. Snow 1933. Experiments on phyllotaxis. II. The effect of displacing a primordium. *Philosophical Transactions of the Royal Society London, Series B* **222**, 353–400.

Stafstrom, J. P. & L. A. Staehelin 1988. Antibody localization of extensin in cell walls of carrot storage roots. *Planta* **174**, 321–32.

Stange, L. 1983. Cell cycle, cell expansion and polarity during morphogenesis of appendicular structures in *Riella helicophylla. Zeitschrift für Pflanzenphysiologie* **112**, 325–35. (Growth of liverwort gemmae)

Steeves, T. A. 1966. On the determination of leaf primordia in ferns. In *Trends in plant morphogenesis*, E. G. Cutter (ed.), 200–219. London: Longman.

Stetler, D. A. & A. E. Demaggio 1972. An ultrastructural study of fern gametophytes during one- to two-dimensional development. *American Journal of Botany* **59**, 1011–17.

Stewart, R. N., F. G. Meyer & H. Dermen 1972. *Camellia* + 'Daisy Eagleson', a graft chimera of *Camellia sasanqua* and *C. japonica. American Journal of Botany* **59**, 515–24.

Sugiyama, M., H. Fukuda & A. Komamine 1986. Effects of nutrient limitation and δ-irradiation on tracheary element differentiation and cell division in single mesophyll cells of *Zinnia elegans. Plant and Cell Physiology* **27**, 601–6.

Sung, Z. R. & R. Okimoto 1983. Coordinate gene expression during somatic embryogenesis in carrots. *Proceedings of the National Academy of Science U.S.A.* **80**, 2661–5. (Two embryo-specific polypeptides)

Sussex, I. M. 1967. Polar growth of *Homosira banksii* zygotes in shake culture. *American Journal of Botany* **54**, 505–10.

Thompson, N. P. 1967. The time course of sieve tube and xylem cell regeneration and their anatomical orientation in *Coleus* stems. *American Journal of Botany* **54**, 588–95. (Xylem differentiates on same side of barrier as phloem)

Thompson, N. P. 1970. The transport of auxin and regeneration of xylem in okra and pea stems. *American Journal of Botany* **57**, 390–93.

Thornley, J. H. M. 1975. Phyllotaxis. I. A mechanistic model. *Annals of Botany* **39**, 491–507.

Thornley, J. H. M. & K. E. Cockshull 1980. A catastrophe model for the switch from vegetative to reproductive growth in the shoot apex. *Annals of Botany* **46**, 333–41.

Tilney-Bassett, R. A. E. 1986. *Plant chimeras*. London: Edward Arnold.

Torrey, J. G. 1963. Cellular patterns in developing roots. *Symposia of the Society for Experimental Biology* **17**, 285–314. (Auxin, root diameter, and vascular pattern)

Tran Thanh Van, K. 1981. Control of morphogenesis in *in vitro* cultures. *Annual Review of Plant Physiology* **32**, 291–311.

Trewavas, A. J. 1986. Resource allocation under poor growth conditions. A major role for growth substances in developmental plasticity. In *Plasticity in plants*, D. H. Jennings & A. J. Trewavas (eds). *Symposia of the Society for Experimental Biology* **40**, 31–76.

Tucker, S. C. 1984. Origin of symmetry in flowers. In *Contemporary problems in plant anatomy*, R. A. White & W. C. Dickson (eds), 351–94. New York: Academic Press. (Asymmetrical development of leguminous flowers)

Tucker, W. Q. J., J. Warren Wilson & P. M. Gresshof 1986. Determination of tracheary element differentiation in lettuce pith explants. *Annals of Botany* **57**, 675–9.

Vaughan, D. 1973. Effects of hydroxyproline on the growth and cell-wall protein metabolism of excised root segments of *Pisum sativum*. *Planta* **115**, 135–45. (Prolongation of cell extension in maturation)

Waaland, S. D. 1984. Positional control of development in algae. In *Positional controls in plant development*, P. W. Barlow & D. J. Carr (eds), 137–56. Cambridge: Cambridge University Press. (*Griffithsia*)

Wada, M., Y. Mineyuki, A. Kadota & M. Furuya 1980. The changes of nuclear position and distribution of circumferentially aligned cortical microtubules during the progression of cell cycle in *Adiantum* protonemata. *Botanical Magazine, Tokyo* **93**, 237–45.

Waisel, Y., I. Noah & A. Fahn 1966. Cambial activity in *Eucalyptus camaldulensis* Dehn. II. The production of phloem and xylem elements. *New Phytologist* **65**, 319–24.

Walbot, V. & C. A. Cullis 1985. Rapid genomic change in higher plants. *Annual Review of Plant Physiology* **36**, 367–96. (Flax genotrophs)

Walker, K. A., M. L. Wendeln & E. G. Jaworski 1979. Organogenesis in callus tissue of *Medicago sativa*. The temporal separation of induction processes from differentiation processes. *Plant Science Letters* **16**, 23–30.

Wardlaw, C. W. 1955. *Embryogenesis in plants*. London: Methuen.

Warren Wilson, J. & P. M. Warren Wilson 1984. Control of tissue patterns in normal development and in regeneration. In *Positional controls in plant development*, P. W. Barlow & D. J. Carr (eds), 235–80. Cambridge: Cambridge University Press. (Cambium regeneration)

Weisenseel, M. H. & R. M. Kicherer 1981. Ionic currents as control mechanisms in cytomorphogenesis. In *Cytomorphogenesis in plants*, O. Kiermayer (ed.), 379–99. Vienna: Springer.

Wetmore, R. H. & J. P. Rier 1963. Experimental induction of vascular tissues in callus of angiosperms. *American Journal of Botany* **50**, 418–30. (Formation of vascular nodules in lilac callus)

Williams, E. G. & G. Maheshwaran 1986. Somatic embryogenesis: factors influencing coordinated behaviour of cells as an embryogenic group. *Annals of Botany* **57**, 443–62. (PEDC and IEDC)

Wochok, Z. S. & I. M. Sussex 1976. Redetermination of cultured root tips to leafy shoots in *Selaginella wildenovii*. *Plant Science Letters* **6**, 185–92. (Root–shoot transformation by low auxin)

Yeung, E. C. 1980. Embryogeny of *Phaseolus*: the role of the suspensor. *Zeitschrift für Pflanzenphysiologie* **96**, 17–28.

Yeung, E. C. & M. E. Clutter 1978. Embryogeny of *Phaseolus coccineus*: growth and microanatomy. *Protoplasma* **94**, 19–40.

Yeung, E. C. & I. M. Sussex 1979. Embryogeny of *Phaseolus coccineus*: the suspensor and the growth of the embryo-proper *in vitro*. *Zeitschrift für Pflanzenphysiologie* **91**, 423–33.

Young, B. S. 1954. The effects of leaf primordia on differentiation in the stem. *New Phytologist* **53**, 445–60. (Auxin replacement of excised leaves in lupin)

REFERENCES

Zeevaart, J. A. D. 1969. *Bryophyllum*. In *The induction of flowering: some case histories*, L. T. Evans (ed.), 435–56. Melbourne: MacMillan.

Zobel, A. M. 1989. Origin of nodes and internodes in plant shoots. I. Transverse zonation of apical parts of the shoot. II. Models of node and internode origin from one layer of cells. *Annals of Botany* **63**, 201–8, 209–20.

Index

Entries in bold type are explained in the Glossary